AMERICAN MATHEMATICAL SOCIETY
TRANSLATIONS
Series 2
Volume 127

Eleven Papers in Analysis

by

P. L. Shabalin
V. A. Zmorovich
A. A. Yakubenko
V. A. Pokhilevich
M. G. Kreĭn
A. A. Nudel′man
G. A. Kurina
G. G. Gevorkyan
A. I. Saginashvili
M. S. Birman
M. Z. Solomyak
L. V. Mikaelyan
V. V. Ivanov
G. N. Kotenkova
M. Ya. Ivanov
V. V. Koretskiĭ
N. Ya. Kurochkina

AMERICAN MATHEMATICAL SOCIETY
PROVIDENCE, RHODE ISLAND

Translation edited by LEV. J. LEIFMAN

1980 *Mathematics Subject Classifications.* Primary 30-XX, 42A38, 42C25, 44A05, 45-XX, 47A10, 47G05, 49B50, 65-XX; Secondary 15A15, 35L65, 35L67, 42A16, 42A63, 46E35, 47B35, 49E25.

Library of Congress Cataloging-in-Publication Data
Main entry under title:

Eleven papers in analysis:
(American Mathematical Society translations; ser. 2, v. 127)
Translations from the Russian.
Table of contents also in Russian.
Includes bibliographies.
1. Mathematical analysis—Addresses, essays, lectures. I. Shabalin, P. L. II. Series.
QA3.A572 ser. 2, vol. 127 510s 85-20126
[QA300.5] [515]
ISBN 0-8218-3094-5

The paper used in this book is acid-free and falls within the guidelines
established to ensure permanence and durability.

Contents

Russian Table of Contents*

---

*The American Mathematical Society scheme for transliteration of Cyrillic may be found at the end of index issues of *Mathematical Reviews*.

В. В. Иванов и Г. Н. Котенкова, Оптимизация алгоритмов по времени их реализации на ЭВМ для решения особых интегральных уравнений, Дифференциальные и интегральные уравнения. Краевые задачи. Тбилисский гос. ун-т, Тбилиси, 1979, 113–126

М. Я. Иванов, В. В. Корецкий и Н. Я. Курочкина, Исследование свойств разностных схем сквозного счета второго порядка аппроксимации, Численные методы динамики сплошной среды. (Новосибирск), **11** (1980), № 2, 41–63

М. Ш. Бирман и М. З. Соломяк, Применение интерполяционных методов к оценкам спектра интегральных операторов, Теория операторов в фунциональных пространствах. Изд-во Наука, Сибирское отделение, Новосибирск, 1977, 42–70

Л. В. Микаелян, Асимптотическое поведение детерминантов Фредгольма усеченных интегральных операторов с матричными ядрами, зависящими от разности аргументов, Доклады Акад. Наук Арм.ССР. **68** (1979), № 3, 137–145

Amer. Math. Soc. Transl.
(2) Vol. **127**, 1986

Univalence Classes and Smirnov Domains*

UDC 517.54

P. L. SHABALIN

Simply-connected single-sheeted domains with rectifiable Jordan boundaries fall into two classes according to whether or not the representation

$$f(\zeta) = \int_0^\zeta \exp \frac{1}{2\pi} \int_{-\pi}^{\pi} S(\zeta e^{-i\theta})\, d\mu(\theta)\, d\zeta, \qquad S(\zeta) = \frac{1+\zeta}{1-\zeta}, \tag{1}$$

of the function f, that maps the disk $E = \{\zeta\colon |\zeta| < 1\}$ conformally on the given domain G, has a singular component. If $\mu(\theta)$ is absolutely continuous, $\mu(\theta) = \mu_a(\theta)$, then $G = f(E)$ is a Smirnov domain (domain of class S) ([1], §14.9). If, however, the measure $d\mu$ has a nontrivial singular component ($d\mu(\theta) = d\mu_a(\theta) + d\mu_s(\theta)$, where $\mu_s(\theta)$ is a nonincreasing singular function), the domain is a non-Smirnov domain. In that case we denote the domain by G_s.

On the other hand, we shall consider classes of domains determined by certain sufficient conditions for the univalence of analytic functions. We shall be interested in the question of whether such classes contain domains of non-Smirnov type. Similar results are of interest from the point of view of the weak problem of univalence [2] of the solutions of inverse boundary value problems. It is known (see [3]) that the general solution of a typical inverse boundary value problem is given by (1). Hence it is useful to know what sufficient conditions for univalence a function of this form, with nontrivial $\mu_s(\theta)$, can satisfy. The point is that domains G and G_s differ not just by local properties of their boundaries. Tumarkin [4] showed that one can obtain a domain of class S from a G_s by an inner deformation of the boundary curve ∂G_s on a set of arbitrarily small linear measure. However, the presence of the singular function $\mu_s(\theta)$ in (1) also affects the multiplicity of the image (see [5]).

1980 *Mathematics Subject Classification*. Primary 30C55, 30E25; Secondary 30C20.
*Translation of Trudy Sem. Kraev. Zadacham No. 16 (1979), 218–226; MR **82c**: 30022.

In the first part of this paper we show that the class of single-sheeted domains defined by a Bazilevich condition [6] contains only Smirnov domains. A Bazilevich condition, $G \in B_{\alpha,\beta}$ is that if $G = f(E)$,

$$\int_{\theta_0}^{\theta} \operatorname{Re}\left\{1 + \zeta\frac{f''(\zeta)}{f'(\zeta)} + (\alpha - 1 + i\beta)\zeta\frac{f'(\zeta)}{f(\zeta)}\right\} d\tau > -\pi, \qquad \zeta = re^{i\tau}, \quad (2)$$

$-\pi < \theta_0 < \theta < \pi$, $\alpha > -\frac{1}{2}$, β a finite real number. We also establish three new criteria for the univalence of analytic functions and investigate the possibility of applying them to mappings on G_s domains.

Let G be a simply-connected single-sheeted domain bounded by a rectifiable curve, and let the analytic function $f(\zeta)$ ($f(0) = 0$, $f'(0) = 1$) map the disk E on G. A sufficient condition for $G \in S$ is that ([1], §15.1)

$$\arg f'(\zeta) > -M, \qquad M = M(f) < \infty, \qquad \zeta \in E. \quad (3)$$

We apply this to prove the following theorem.

THEOREM 1. *If G satisfies a Bazilevich condition, then $G \in S$.*

PROOF. Choose θ_0 so that $\arg f'(re^{i\theta_0}) = 0$, $0 < r < 1$; we can transform (2) into the form

$$\arg f'(re^{i\theta}) > -\pi - (\theta - \theta_0) - |\alpha - 1 + i\beta| \int_{-\pi}^{\pi} \left|\zeta\frac{f'(\zeta)}{f(\zeta)}\right| d\tau.$$

Let us show that $\int_{-\pi}^{\pi} |\zeta f'(\zeta)/f(\zeta)|\, d\tau$ is uniformly bounded in E.

In fact, by the Koebe covering theorem we have $|f(e^{i\theta})| > \frac{1}{4}$ at every point of the boundary of the unit disk E. Since the function $f(\zeta)/\zeta$ is regular in E and never 0, we have $|f(\zeta)/\zeta| > \frac{1}{4}$ throughout E by the minimum principle. Hence, by using the rectifiability of ∂G, we obtain the inequality

$$\int_{-\pi}^{\pi} \left|\zeta\frac{f'(\zeta)}{f(\zeta)}\right| d\tau < 4\int_{-\pi}^{\pi} |f'(\zeta)|\, d\tau < 4l, \qquad \zeta = re^{i\tau}, 0 < r < 1,$$

where l is the length of ∂G.

The preceding inequality shows that (3) is satisfied with

$$M = 3\pi + 4l|\alpha - 1 + i\beta|.$$

Thus a Bazilevich class contains only Smirnov domains. As was shown in [7] and [8], non-Smirnov domains are contained in the classes generated by the univalence conditions of Nehari and Becker. We shall extend the Becker condition in the following way. We do not require that $|\zeta f''(\zeta)/f'(\zeta)| < (1 - |\zeta|^2)^{-1}$ in the whole disk E, but only in an annulus $E[q, 1) = \{\zeta\colon\ q \leqslant |\zeta| < 1\}$. A function $f(\zeta)$ that satisfies this condition is finitely valent. We ask for a supplementary condition on $f(\zeta)$ in the rest of $E_q = \{\zeta\colon\ |\zeta| < q\}$ that will guarantee univalence. The example of the function $f(\zeta) = \exp\{a\zeta\}$ with $a > \pi$ shows that, generally speaking, the mere univalence of $f(\zeta)$ in E_q is not sufficient. Hence we shall suppose that, in addition to the Becker inequality in $E[q, 1)$, $f(\zeta)$ satisfies some analytic conditions in E_q. We have the following theorem.

THEOREM 2. *If $f(\zeta)$ is analytic in E, satisfies $f(0) = 0$ and $f'(0) = 1$, and also*

$$|\arg f'(\zeta)| < \arccos q, \qquad \zeta \in \bar{E}_q, \tag{4}$$

and

$$\left|\frac{f''(\zeta)}{f'(\zeta)}\right| < \frac{1}{1-|\zeta|^2}, \qquad \zeta \in E[q,1), \tag{5}$$

then $f(\zeta)$ is univalent in E.

PROOF. It is easily seen that the function $g(\zeta) = f(\rho\zeta)$, $0 < \rho < 1$, is analytic in the closed disk $\bar{E}$ and satisfies

$$\left|\frac{g''(\zeta)}{g'(\zeta)}\right| < \frac{\rho}{1-(\rho|\zeta|)^2}, \qquad \zeta \in E[q,1). \tag{5'}$$

Then the function

$$\tilde{g}(\zeta) \rightleftharpoons \begin{cases} g(\zeta), & \zeta \in E, \\ g_1(\zeta), & \zeta \in E(1,1/q), \end{cases}$$

where $g_1(\zeta) = g(1/\bar{\zeta}) + (\zeta - 1/\bar{\zeta})g'(1/\bar{\zeta})$, maps $E_{1/q}$ $(1+\rho)/(1-\rho)$-quasiconformally on the domain $\tilde{D}$, where

$$\partial\tilde{D} = \{z\colon z = g(qe^{i\theta}) + (q^{-2}-1)qe^{i\theta}g'(qe^{i\theta}),\ \theta \in [-\pi,\pi]\}.$$

In fact, $\tilde{g}(\zeta)$ is continuous in $\bar{E}_{1/q}$, the complex deviation of $\mu(\zeta)$ is everywhere 0 for $\zeta \in \bar{E}$, and by (5′),

$$|\mu(\zeta)| = \left|\frac{\tilde{g}_{1\bar{\zeta}}}{\tilde{g}_{1\zeta}}\right| = \left|\frac{1}{\bar{\zeta}}\left(1-\frac{1}{|\zeta|^2}\right)\frac{g''(1/\bar{\zeta})}{g'(1/\bar{\zeta})}\right| < \rho$$

for $\zeta \in E(1,1/q]$. Since $\tilde{g}(\zeta)$ is locally univalent in $\bar{E}_{1/q}$ (since $I = |\tilde{g}_\zeta|^2 - |\tilde{g}_{\bar{\zeta}}|^2 = |\tilde{g}_\zeta|^2(1-|\mu(\zeta)|^2) > 0$), the number $n(a)$ of a-points of $\tilde{g}$ is given by the formula ([9], p. 71)

$$2n(a) = \gamma + (2-\nu) + 2q(a) - p. \tag{6}$$

Here γ is the number of branch points, ν is the connectivity of the domain, $q(a)$ is the sum of the orders of the image of the boundary circle with respect to the point $z = a$, and p is the sum of the angular orders of the boundary circle.

We study the behavior of the curve $\partial\tilde{D}$, which may be considered as the image of the circle $|\zeta| = q$ under the analytic mapping

$$\varphi(\zeta) = g(\zeta) + (q^{-2}-1)\zeta g'(\zeta).$$

It is therefore a smooth closed curve. Let us show that $\partial\tilde{D}$ has no self-intersections. In fact, the harmonic function $\operatorname{Re}\varphi'(\zeta)$ has the representation

$$\operatorname{Re}\varphi'(\zeta) = \operatorname{Re}\{g'(\zeta) + (1-q^2)\zeta g''(\zeta)\}q^{-2}, \qquad \zeta \in \bar{E}_q.$$

By the maximum principle we have

$$\operatorname{Re}\varphi'(\zeta) > \{\operatorname{Re} g'(\zeta_0) - (1-q^2)|\zeta_0 g''(\zeta_0)|\}q^{-2}, \qquad \zeta_0 = qe^{i\varphi_0}.$$

If we estimate $g''(\zeta_0)$ by using (5′), we obtain

$$q^2 \operatorname{Re} \varphi'(\zeta) > \operatorname{Re} g'(\zeta_0) - q|g'(\zeta_0)|, \qquad \zeta \in \overline{E}_q.$$

Now the inequality $\operatorname{Re}\varphi'(\zeta) > 0$, which guarantees the univalence of $\varphi(\zeta)$ in the closed disk $\overline{E}_q$, follows from (4). Thus $\partial\tilde{D}$ is a simple curve which goes once around the origin.

We now turn to (6). For a regular curve, p is the number of revolutions of the tangent when $\partial\tilde{D}$ is traversed in the positive sense, and $q(a)$ is the number of rotations of the gradient vector of g when its tail traverses $\partial\tilde{D}$. In the present case $p = q(a) = 1$, the number of preimages of the branch points is $\gamma = 0$, and consequently if $a \in \tilde{D}$ the equation $\tilde{g}(\zeta) = a$ has exactly one solution.

Thus, we have shown that $g(\zeta)$ is univalent in the closed disk $\overline{E}$. Letting ρ tend to 1, we obtain the conclusion of the theorem.

The following theorem can be proved by the same method.

THEOREM 3. *If $f(\zeta)$ is analytic in E and satisfies $f(0) = 0$, $f'(0) = 1$, and the conditions*

$$\left|\arg\left\{\zeta\frac{f'(\zeta)}{f(\zeta)}\right\}\right| < \frac{\pi}{2} - \arcsin q, \qquad \zeta \in \overline{E}_q,$$
$$\left|\frac{f''(\zeta)}{f'(\zeta)}\right| < \frac{1}{1-|\zeta|^2}, \qquad \zeta \in E[q,1), \tag{7}$$

then $f(\zeta)$ is univalent in E.

The proof differs from the proof given above only in the grounds for the simplicity of the curve $\partial\tilde{D}$. We again consider $\partial\tilde{D}$ as the image of the circle $|\zeta| = q$ under the mapping by the analytic function $\varphi(\zeta)$. We represent the harmonic function $\arg\{\zeta\varphi'(\zeta)/\varphi(\zeta)\}$ in E_q in the form

$$\arg\left\{\zeta\frac{\varphi'(\zeta)}{\varphi(\zeta)}\right\} = \arg\left\{1 + (1-q^2)\zeta\frac{g''(\zeta)}{g'(\zeta)}\right\} - \arg\left\{1 - q^2 + q^2\frac{g(\zeta)}{\zeta g'(\zeta)}\right\}.$$

Hence we obtain the inequality

$$\left|\arg\left\{\zeta\frac{\varphi'(\zeta)}{\varphi(\zeta)}\right\}\right| \leqslant \left|\arg\left\{1 + (1-q^2)\zeta_0\frac{g''(\zeta_0)}{g'(\zeta_0)}\right\}\right| + \left|\arg\left\{1 - q^2 + q^2\frac{g(\zeta_0)}{\zeta_0 g'(\zeta_0)}\right\}\right|.$$

Since by (5), written for the circle $|\zeta| = q$, we have

$$\left|\arg\left\{1 + (1-q^2)\zeta_0\frac{g''(\zeta_0)}{g'(\zeta_0)}\right\}\right| < \arcsin q, \qquad \zeta_0 = qe^{i\varphi_0},$$

then the evident inequality

$$\left|\arg\left\{1-q^2+q^2\frac{g(\zeta_0)}{\zeta_0 g'(\zeta_0)}\right\}\right| \leqslant \left|\arg\left\{\zeta_0\frac{g'(\zeta_0)}{g(\zeta_0)}\right\}\right|$$

and (7) establish the univalence (and star-shaped character) of $\varphi(\zeta)$ in the closed disk $\overline{E}_q$.

Similarly we can prove the following theorem:

THEOREM 4. *If the analytic function $f(\zeta)$ satisfies the conditions*

$$\operatorname{Re}\zeta\frac{f''(\zeta)}{f'(\zeta)} > -1, \qquad \zeta \in \overline{E}_q,$$

$$\left|\zeta\frac{f''(\zeta)}{f'(\zeta)}\right| < \frac{1}{1-|\zeta|^2}, \qquad \zeta \in E[q,1),$$

then $f(\zeta)$ is univalent in E.

Here the simplicity of the curve $\partial\tilde{D}$ is a consequence of the known almost convexity of the function

$$\varphi(\zeta) = g(\zeta) + (q^{-2}-1)g'(\zeta).$$

We note that Theorem 4 was in effect proved by Avkhadiev (see [10]) by a somewhat different method. Namely, [10] gave a construction of a continuation of $g(\zeta)$ from the disk $E_{1/q}$ to the whole complex plane and established the possibility of applying Hadamard's theorem. The convexity of the image of E_q was not used, but only the convexity of a single curve, the image of the circle $|\zeta| = q$. This allowed the theorem to be stated for the annulus $E[q,1)$.

We now consider the possibility of applying our conditions for univalence to mappings of the disk E onto domains that do not belong to the Smirnov class. We form the function

$$F(\zeta) = \int_0^\zeta \exp\{k\ln f'(\zeta)\}\,d\zeta, \qquad \zeta \in E, \tag{8}$$

where $f(\zeta)$ is defined by (1) and k is a positive constant. In the representation (1), let $\mu(\theta)$ have a nontrivial singular component $\mu_s(\theta)$. Then (8) is a mapping of the unit disk on a domain $G_s \notin S$.

Let Λ be the class of real functions that satisfy Zygmund's condition: $\mu(\theta+h) - 2\mu(\theta) + \mu(\theta-h) = O(h)$ uniformly in $\theta \in (-\pi,\pi)$, with $\mu(\theta)$ continuous. Zygmund's class of almost-smooth functions contains singular functions $\mu_s(\theta)$, as was shown constructively by Piranian [11]. We prove the following result:

ASSERTION 1. *There is a positive number k such that if $\mu(\theta) \in \Lambda$, the function* (8) *satisfies the hypotheses of Theorem* 4.

PROOF. We may suppose without loss of generality that $\mu(-\pi) = \mu(\pi)$. Then (8) may be rewritten in the form

$$\ln F'(\zeta) = \frac{\partial}{\partial\varphi}\left\{\frac{k}{2\pi}\int_{-\pi}^{\pi}\mu(\theta)S(\zeta e^{-i\theta})\,d\theta\right\}, \qquad \zeta = re^{i\varphi}.$$

If we let $h(\zeta)$ denote the function $\frac{1}{2\pi}\int_{-\pi}^{\pi}\mu(\theta)S(\zeta e^{-i\theta})\,d\theta$, the preceding equality implies that

$$i\zeta\frac{F''(\zeta)}{F'(\zeta)} = k\frac{\partial^2}{\partial\varphi^2}h(\zeta), \qquad \zeta = re^{i\varphi}.$$

Now $h(\zeta)$ is regular in E and continuous in $\overline{E}$, $\operatorname{Re} h(e^{i\theta}) = \mu(\theta)$, and $\mu(\theta) \in \Lambda$. Then by a theorem of Zygmund ([12], p. 109) we have the estimate

$$\left|\frac{\partial^2}{\partial\varphi^2}h(\zeta)\right| = o\{(1-|\zeta|)^{-1}\},$$

which evidently implies that

$$\left|\zeta\frac{F''(\zeta)}{F'(\zeta)}\right| = ko\{(1-|\zeta|)^{-1}\}. \tag{9}$$

Since the function $\zeta F''(\zeta)/F'(\zeta)$ is regular in E, we have by (9)

$$\left|\zeta\frac{F''(\zeta)}{F'(\zeta)}\right| < \frac{kM(F)}{1-|\zeta|^2}, \qquad \zeta \in [q,1),\, 0 < q < 1.$$

Then the hypotheses of Theorem 4 are satisfied provided that $k < (1-q^2)/M(F)$.

We consider the univalence condition of Theorem 2 from this point of view.

ASSERTION 2. *There is a positive number k such that if $\mu(\theta) \in \Lambda$, the function* (8) *satisfies conditions* (4) *and* (5).

PROOF. From (8) we obtain

$$\arg F'(\zeta) = \frac{k}{2\pi}\int_{-\pi}^{\pi} Q(r, \theta-\varphi)\,d\mu(\theta),$$

where

$$Q(r,\theta-\varphi) = \operatorname{Im} S(\zeta e^{-i\theta}) = \frac{2r\sin(\theta-\varphi)}{1+r^2-2r\cos(\theta-\varphi)}, \qquad \zeta = re^{i\varphi}.$$

After integrating by parts, supposing without loss of generality that $\mu(-\pi) = \mu(\pi)$, we have

$$\arg F'(\zeta) = \frac{k}{2\pi}\int_{-\pi}^{\pi} \mu(\theta)\frac{\partial}{\partial\theta}Q(r,\theta-\varphi)\,d\theta,$$

where

$$\frac{\partial}{\partial\theta}Q(r,\theta-\varphi) = \frac{2r[(1+r^2)\cos(\theta-\varphi)]}{[1+r^2-2r\cos(\theta-\varphi)]^2}.$$

Since the function $\partial Q(r,\theta-\varphi)/\partial\theta$ is even, we can transform the preceding expression to

$$\arg F'(\zeta) = \frac{k}{2\pi}\int_{-\pi}^{\pi}\frac{1}{2}[\mu(\varphi+t) - 2\mu(\varphi) + \mu(\varphi-t)]\frac{\partial}{\partial t}Q(r,t)\,dt.$$

Since $\partial Q(r,t)/\partial t$ changes sign at the point $t = \tau(r) = \arccos\{2r/(1+r^2)\}$, we have

$$|\arg F'(\zeta)| < \frac{k}{2\pi}\left\{\left(\int_0^\tau - \int_\tau^\pi\right) O(t)\frac{\partial}{\partial t} Q(r,t)\,dt\right\}.$$

If we transform the preceding inequality into

$$|\arg F'(\zeta)| < \frac{k}{2\pi}\left\{O(\tau)Q(r,\tau) + \left(\int_0^\tau - \int_\tau^\pi\right) Q(r,t)\,dt\right\}$$

and take account of the easily verified relations

$$\tau = O(1-r), \qquad Q(r,\tau) = O\{(1-r)^{-1}\},$$

$$\int Q(r,t)\,dt = O\left\{\ln\left(\frac{1}{1-r}\right)\right\},$$

we finally obtain

$$|\arg F'(\zeta)| = kO\left\{\ln\frac{1}{1-r}\right\}.$$

This means that $|\arg F'(\zeta)| < kN(F)\ln(1/(1-|\zeta|))$ for all $\zeta \in E$. Since, moreover, the inequality

$$\left|\frac{F''(\zeta)}{F'(\zeta)}\right| < \frac{kM(F)}{1-|\zeta|^2}, \qquad \zeta \in E,$$

follows from (9), in order to obtain hypotheses (4) and (5) of Theorem 2 it is enough to take

$$k = \min\{1/M(F);\ \arccos q/N(F)\ln(1-q)^{-1}\}.$$

The conditions for univalence given in Theorems 2–4 allow us to draw some conclusions about the geometric properties of mappings of the open disk E. For example, Assertion 1 shows that there is a mapping onto a G_s domain (that is, onto a domain whose boundary is extremely unsmooth), carrying a circle concentric with the unit circle and arbitrarily close to it onto a convex curve.

The author thanks Professor L. A. Aksent′ev for supervision of this work and for helpful advice.

REFERENCES

1. I. I. Privalov, *Boundary properties of analytic functions*, GITTL, Moscow, 1950; German transl., VEB Deutscher Verlag Wiss., Berlin, 1956.

2. F. G. Avkhadiev and L. A. Aksent′ev, *Fundamental results on sufficient conditions for the univalence of analytic functions*, Uspekhi Mat. Nauk **30** (1975), no. 4 (184), 3–60; English transl. in Russian Math. Surveys **30** (1975).

3. F. D. Gakhov, *On inverse boundary value problems*, Kazan. Gos. Univ. Uchen. Zap. 113 (1953), no. 10, 9–20. (Russian)

4. G. Ts. Tumarkin, *The influence of a deformation of the boundary on certain conformal properties of domains*, Mat. Zametki **8** (1970), 351–359; English transl. in Math. Notes **8** (1970).

5. L. A. Aksent′ev and P. L. Shabalin, *On the problems of S. N. Andrianov*, Trudy Sem. Kraev. Zadacham No. 14, (1977), 28–35. (Russian)

6. F. G. Avkhadiev and L. A. Aksent′ev, *Functions of Bazilevich class in the disk and annulus*, Dokl. Akad. Nauk SSSR **214** (1974), 241–244; English transl. in Soviet Math. Dokl. **15** (1974).

7. P. L. Duren, H. S. Shapiro and A. L. Shields, *Singular measures and domains not of Smirnov type*, Duke Math. J. **33** (1966), 247–254.

8. P. L. Shabalin, *On the univalence of the general solution of the interior inverse boundary value problem*, Izv. Vyssh. Uchebn. Zaved. Matematika **1975**, no. 12 (163), 92–95; English transl. in Soviet Math. (Iz. VUZ) **19** (1975).

9. Marston Morse, *Topological methods in the theory of functions of a complex variable*, Princeton Univ. Press, Princeton, N. J., 1947.

10. F. G. Avkhadiev and P. L. Shabalin, *On mappings onto multiply connected domains that do not belong to the Smirnov class*, Manuscript No. 2550-76, deposited at VINITI, 1976. (Russian) R. Zh. Mat. **1976** #11Б157.

11. George Piranian, *Two monotonic, singular, uniformly almost smooth functions*, Duke Math. J. **33** (1966), 255–262.

12. A. Zygmund, *Trigonometric series*, 2nd rev. ed., Vol. I, Cambridge Univ. Press, 1959.

Translated by R. P. BOAS

Amer. Math. Soc. Transl.
(2) Vol. **127**, 1986

On the Radius of α-Convexity for Order γ in the Class $S_\beta^*(m)$ †

UDC 517.53

V. A. ZMOROVICH AND A. A. YAKUBENKO

In this note we solve the problem of the radius $r_0 \in]0, 1[$ of the largest disk with center at $z = 0$ in which every function $f(z)$ of the class $S_\beta^*(m)$ (the class of m-fold symmetric (m an integer > 1) star-like functions of order β ($0 \leqslant \beta < 1$) in $E(z\colon |z| < 1)$) satisfies $\operatorname{Re} I(f; \alpha) > \gamma$, where

$$I(f;\alpha) = (1-\alpha)\frac{zf'}{f} + \alpha\left(1 + \frac{zf''}{f'}\right), \qquad \alpha \neq 0, \quad \alpha \in]-\infty, +\infty[, \quad \gamma < 1. \tag{1}$$

The problem is solved completely for the cases 1) $\alpha > 0$ and 2) $\alpha < 0$, $\beta \geqslant \sqrt{2} - 1$. When $\alpha < 0$ and $0 \leqslant \beta < \sqrt{2} - 1$, the problem is solved under the supplementary condition $m|\alpha| \geqslant 2(1 - \beta)$.

This result contains as special cases the results of [1], which were the most general among published results (they are listed in [1]).

We have established two theorems.

THEOREM 1. *Let* $h = \beta/(1-\beta)$, $0 \leqslant \beta < 1$, $\alpha > 0$, $k = [m\alpha(1+h)]^{-1}$, $m \geqslant 1$ *an integer*, $A = 1 + k$, $\gamma_1 = \gamma/m\alpha$, $\gamma < 1$, $B = kh^2 - \gamma_1 h$ *and* $C = 2h(1-h) + 2hk - h\gamma_1$.

If $kh - \gamma_1 \leqslant 0$ (1), *then*

$$r_0 = \sqrt[m]{|\eta_0 - 1|/(\eta_0 + 1)}, \tag{2}$$

where η_0 *is the smallest positive root of the equation*

$$\Phi(\eta) \equiv (2A - 1)\eta^2 + 2C\eta + 2B - 1 = 0. \tag{3}$$

If $kh - \gamma_1 > 0$, *then when* $\Phi(\sqrt{B/A}) \geqslant 0$ *we must take* $\eta_0 = \sqrt{B/A}$ *in* (2). *If* $\Phi(\sqrt{B/A}) < 0$, *the number* η_0 *in* (2) *is the least positive root of* (3).

1980 *Mathematics Subject Classification*. Primary 30C45.
† Translation of Dokl. Akad. Nauk Ukrain. SSR Ser. A **1981**, no. 10, 6–9; MR **83h**: 30015.

THEOREM 2. *Let $\alpha < 0$ and $|\alpha| = \delta$, with h, k and m the same as in the hypotheses of Theorem* 1. *Set*

$$A_1 = [m\delta(1+h)]^{-1}, \qquad B_1 = [h^2 + m\delta(1+h) - \gamma h(1+h)]A_1,$$
$$C_1 = [2h - \gamma(1+h)]A_1$$

and suppose that $h \geqslant 1/\sqrt{2}$, or $0 \leqslant h < 1/\sqrt{2}$, but now $m\delta \geqslant 2(1-\beta)$. Then if $B_1 \leqslant 0$, the number r_0 is defined by (2), *where η_0 is the positive root of the equation*

$$\Phi_1(\eta) \equiv (2A_1 - 1)\eta^2 + 2C_1\eta + 2B_1 - 1 = 0. \tag{4}$$

If $B_1 > 0$ and $\Phi_1(\sqrt{B_1/A_1}) \geqslant 0$, we must take $\eta_0 = \sqrt{B_1/A_1}$ in (2); *if $\Phi_1(\sqrt{B_1/A_1}) < 0$, then η_0 in* (2) *is to be taken to be the least positive root of equation* (4).

We omit the explicit expressions for r_0 because of their complexity.

For the proofs of both theorems we start from the fact that $f \in S_\beta^*(m)$ if and only if

$$\frac{zf'}{f} = \frac{p+h}{1+h}, \tag{5}$$

where $p \equiv p(z) \in \mathscr{P}(m)$, i.e. $p(z)$ is regular in E, $p(0) = 1$, $\operatorname{Re} p(z) > 0$ in E, and $p(ze^{2\pi i/m}) = p(z)$, $z \in E$. From (5), we have

$$I(f;\alpha) = \frac{p+h}{1+h} + \alpha\frac{zp}{p+h}.$$

Hence if we apply the main theorem of [2] we obtain

$$\operatorname{Re} I(f;\alpha) > \operatorname{Re}\left(\frac{p+h}{1+h} + \frac{m\alpha}{2}\cdot\frac{p^2-1}{p+h}\right) - \frac{m|\alpha|}{2}\cdot\frac{\rho^2 - |p-a|^2}{|p+h|}, \tag{6}$$

where

$$a = \frac{1+r^{2m}}{1-r^{2m}}, \quad \rho = \frac{2r^m}{1-r^{2m}}, \quad |z| = r, \quad \rho \geqslant |p-a|.$$

It is known that (6) is sharp, with equality attained for

$$p(z) = \lambda\frac{1+z^{2m}e^{-i\varphi_1}}{1-z^{2m}e^{-i\varphi_1}} + (1-\lambda)\frac{1+z^{2m}e^{-i\varphi_2}}{1-z^{2m}e^{-i\varphi_2}},$$

where $\lambda \in [0,1]$, $\varphi_j \in [0,2\pi]$, and $j = 1,2$. If we put $p = a + x + iy$ in (6), where $x^2 + y^2 \leqslant \rho^2$, then, for $\alpha > 0$,

$$\operatorname{Re} I(f;\alpha) \geqslant F(x,y^2) \equiv \frac{x+a+h}{1+h} + \frac{m\alpha}{2}\cdot(x+a+h)$$
$$+ \frac{m\alpha}{2}\cdot\frac{(h^2-1)(x+a+h)}{(x+a+h)^2+y^2} + \frac{m\alpha}{2}\cdot\frac{x^2+y^2-\rho^2}{\sqrt{(x+a+h)^2+y^2}} - m\alpha h.$$

If we calculate $\partial F/\partial(y^2)$ we see easily that $\partial F/\partial(y^2) > 0$ for $x \in [-\rho,\rho]$. Therefore the absolute minimum of $F(x,y^2)$ in the disk $x^2 + y^2 \leqslant \rho^2$ is attained on the diameter $y = 0$, $-\rho \leqslant x \leqslant \rho$.

Putting $a + x + h = R$, we have

$$F(x,0) \equiv F_1(R) = \left(\frac{1}{1+h} + m\alpha\right)R + \frac{m\alpha(a+h)h}{R} - (2h+a)m\alpha,$$

$$a - \rho + h \leqslant R \leqslant a + \rho + h, \qquad \rho = \sqrt{a^2-1}\,.$$

Consider the equation $F_1(R) = \gamma$. We look for the smallest value a_0 of $a > 1$ for which this equation has a root $R \in [a-\rho+h,\ a+\rho+h]$. Since $-\rho \leqslant R - a - h \leqslant \rho$, we have $(R-a-h)^2 \leqslant \rho^2$, i.e. $1 \leqslant (R-h)(2a+h-R)$.

If we put $R - h = \eta$, we obtain

$$1 \leqslant \eta(2a - \eta). \tag{7}$$

If we consider a and η as Cartesian coordinates in the plane, we see easily that (7) implies that the point (a, η) lies in the closed domain D which is bounded on the right by the branch of the hyperbola

$$\eta(2a-\eta) = 1, \tag{8}$$

whose asymptotes are $\eta = 0$ and $\eta = 2a$.

The equation $F_1(R) = \gamma$ can be put in the form

$$a = A\eta + B/\eta + C, \tag{9}$$

where A, B and C are as specified in Theorem 1. The curve (9) is also a hyperbola if $B \neq 0$. From (8) we have $da/d\eta = A - B/\eta^2$. If $B \leqslant 0$ we have $da/d\eta > 0$. Therefore the smallest value $a = a_0$ is given by $a_0 = \frac{1}{2}(\eta_0 + 1/\eta_0)$, where η_0 is the smallest positive root of (3). Since

$$r_0 = \sqrt[2m]{(a_0-1)/(a_0+1)}$$

we obtain (2).

If $B > 0$, the absolute minimum of a is attained for $\eta_0 = \sqrt{B/A}$ and equals $a_0 = A\eta_0 + B/\eta_0 + C$. A necessary and sufficient condition for the point (a_0, η_0) to lie in D is that $\eta_0(2a - \eta_0) \geqslant 1$, i.e. $\Phi(\sqrt{B/A}) \geqslant 0$. If this condition is not satisfied, we have to find the intersection of the hyperbolas (8) and (9), which again leads us to having to determine the smallest positive root of (3). This completes the proof of Theorem 1.

The same "two hyperbolas" method can also be applied to prove Theorem 2, if we first show that the absolute minimum of the right-hand side of (6) is attained on the diameter $y = 0$, $-\rho \leqslant x \leqslant \rho$. However, when $\alpha < 0$ we have succeeded in doing this only in the two cases given in the hypothesis of Theorem 2.

We do not consider it expedient to give the details of the proof of Theorem 2, since the reasoning is similar to that in the proof of Theorem 1. As for the case $0 \leqslant h < 1/\sqrt{2}$ when $m|\alpha| < 2/(1+h)$, this presents significant difficulties and will require a change in the method of investigation.

References

1. P. I. Sizhuk and V. V. Chernikov, *The radius of α-convexity of order β for the class of starshaped functions of order γ*, Extremal Problems of Function Theory, Izdat. Tomsk. Univ., Tomsk, 1979, pp. 49–58. (Russian)

2. V. A. Zmorovich, *On some theorems in the theory of extremal estimates for special classes of analytic functions*, Dopovidī Akad. Nauk Ukraïn. RSR Ser. A **1965**, 980–983. (Russian)

Translated by R. P. BOAS

Amer. Math. Soc. Transl.
(2) Vol. **127**, 1986

On the Order of Starlikeness of the Class of α-Convex Functions of Order γ*

UDC 517.54

V. A. ZMOROVICH AND V. A. POKHILEVICH

Let $\alpha \in]-\infty, +\infty[$, $0 \leqslant \gamma < 1$. A function $F(z) = z + a_2 z^2 + \cdots$, regular in the disk $E = \{z, |z| < 1\}$ is said to be *of class* $M_\alpha(\gamma)$ if $F(z)$ satisfies, in E, the conditions a) $F(z) \cdot F'(z)/z \neq 0$; b) $\operatorname{Re} J(F, \alpha) \geqslant \gamma$, where

$$J(F, \alpha) = (1 - \alpha)\frac{zF'(z)}{F(z)} + \alpha\left(1 + \frac{zF''(z)}{F'(z)}\right).$$

The class $M_\alpha(\gamma)$ is evidently a subclass of Mocanu's class of α-convex functions [1], [2], and coincides with it when $\gamma = 0$. It follows from the results of [1]–[4] that $M_\alpha(\gamma) \subset S^*$, where S^* is the class of functions that are regular, univalent, and starlike in E. We recall that a function $F(z)$ that is regular and univalent in E is *starlike of order* β, $\beta \in (0, 1)$, if and only if it satisfies the condition $\operatorname{Re}(zF'(z)/F(z)) > \beta$ in E.

Up to now there has not been a complete solution of the problem of the exact value of the quantity

$$\beta(\alpha, \gamma) = \inf_{F \in M_\alpha} \inf_{z \in E} \operatorname{Re}\frac{zF'(z)}{F(z)},$$

which is called the *order of starlikeness* of the class $M_\alpha(\gamma)$.

For $\alpha \leqslant 0$ and $0 \leqslant \gamma < 1$, it has been shown [5], [6] that $\beta(\alpha, \gamma) = 0$. It was shown in [7] that $\beta(\alpha, 0) = 0$ for $0 < \alpha < 1$.

In this note we present a theorem that completely solves the problem.

1980 *Mathematics Subject Classification*. Primary 30C45.
* Translation of Dokl. Akad. Nauk Ukrain. SSR Ser. A **1982**, no. 9, 5–7; MR **84f**: 30026.

THEOREM. *The order of starlikeness* $\beta(\alpha, \gamma)$ *for the class* $M_\alpha(\gamma)$ *is given by the formula*

$$\beta(\alpha,\gamma) = \begin{cases} \dfrac{\gamma(2-\sigma)}{2(1+\gamma-\sigma)} \equiv \beta_1 & \text{for } \sigma < 1; \quad (1) \\ 1/2 & \text{for } \sigma = 1; \quad (2) \\ \left[\displaystyle\int_0^1 \left(\frac{2}{1+t^\alpha}\right)^\nu dt\right]^{-1} \equiv \beta_2 & \text{for } \sigma > 1, \quad (3) \end{cases}$$

where $\sigma = \alpha + 2\gamma$ *and* $\nu = 2(1-\gamma)/\alpha$. *The inequalities* (1)–(3) *are sharp, and attained by the function* $F_0(z) \in M_\alpha(\gamma)$ *such that*

$$\frac{zF_0'(z)}{F_0(z)} = \left[(1+z)^\nu \int_0^1 \frac{dt}{(1+zt^\alpha)^\nu}\right]^{-1} \overset{\text{def}}{=} w_0(z).$$

It is evident that w_0 belongs to the Carathéodory class $\mathscr{P}$, i.e. the class of functions that are regular in E and satisfy $p(0) = 1$ and $\operatorname{Re} p(z) > 0$.

REMARK. For $\alpha + 2\gamma = 1$ ($\sigma = 1$), formulas (1) and (3) coincide, so that the transition from one formula to another is continuous except in the case $\gamma = 0$ investigated in [7].

The proof of the theorem is based on an analysis of the regular solution of the differential equation

$$w + \frac{\alpha z w'}{w} = \frac{p+h}{1+h}, \tag{5}$$

where $p \in \mathscr{P}$, $h = \gamma/(1-\gamma)$, and $\gamma \in [0, 1)$. The function $w_0(z)$ corresponding to the extremal function $F_0(z)$ is obtained from (5) for $p(z) = (1-z)/(1+z)$, as follows from the results of [8] and [9]. For $\sigma \geqslant 2$, $w_0(z)$ is studied by the method of subordinate functions, which leads easily to the inequality

$$\left|\frac{1}{w_0(z)} - \frac{1}{2\beta_2}\right| \leqslant \frac{1}{2\beta_2},$$

which is equivalent to $\operatorname{Re} w_0(z) \geqslant \beta_2$. For $\sigma = 1$ the number $\beta(\alpha, \gamma)$ can be found directly from (4). Finally, when $0 < \sigma < 1$ or $1 < \sigma < 2$ it is expedient to simplify the analysis by introducing the functions

$$u(\varphi) = \operatorname{Re}\frac{1}{w_0(e^{i\varphi})}; \qquad v(\varphi) = \operatorname{Im}\frac{1}{w_0(e^{i\varphi})}$$

($0 \leqslant \varphi < \pi$), which are solutions of the system of ordinary differential equations

$$\alpha u'(\varphi) + (1-\gamma)\operatorname{tg}(\varphi/2)u(\varphi) - \gamma v(\varphi) = 0,$$
$$\alpha v'(\varphi) + (1-\gamma)\operatorname{tg}(\varphi/2)v(\varphi) + \gamma u(\varphi) = 1$$

with the initial conditions $u(0) = 1/\beta_2$ and $v(0) = 0$.

The details of the analysis are rather cumbersome and frequently elementary, and therefore we omit them. In conclusion we present a table of $\beta(\alpha, \gamma)$ for certain values of α and γ. The number $\beta(\alpha, \gamma)$ appears at the intersection of the corresponding row and column. For $\alpha + 2\gamma > 1$ the results were calculated on an Èlektronika B3–34 computer.

	α								
γ	0.4	0.5	0.6	0.7	0.8	0.9	1.1	1.2	1.3
0.2	0.30	0.3(6)	0.5	0.523	0.5444	0.5638	0.5988	0.6152	0.6309
0.3	0.5	0.5276	0.5517	0.5730	0.5921	0.6096	0.6415	0.6565	0.6711
0.4	0.5615	0.5857	0.6068	0.6256	0.6425	0.6580	0.6864	0.6999	0.7131
0.5	0.6267	0.6471	0.6651	0.6811	0.6955	0.7088	0.7334	0.7453	0.7570
0.6	0.6954	0.7118	0.7264	0.7394	0.7512	0.774	0.7826	0.7227	0.8028
0.7	0.7673	0.7766	0.7907	0.8005	0.8095	0.8179	0.8340	0.8421	0.8505
0.8	0.8423	0.8504	0.8577	0.8634	0.8704	0.8761	0.8875	0.8936	0.9001
0.9	0.9115	0.9186	0.9275	0.9308	0.9337	0.9366	0.9432	0.9432	0.9517

Values of $\beta(\alpha, \gamma)$ for certain values of α and γ

Summary. A new result has been obtained—the order of starlikeness has been found for the class of α-convex functions of order γ, depending on α and γ, $\alpha \in [0, +\infty)$, $\gamma \in [0, 1)$.

REFERENCES

1. Petru T. Mocanu, *Une propriété de convexité généralisée dans la théorie de la représentation conforme*, Mathematica (Cluj) **11 (34)** (1969), 127–133.

2. Petru T. Mocanu and Maxwell O. Reade, *On generalized convexity in conformal mappings*, Rev. Roumaine Math. Pures Appl. **16** (1971), 1541–1544.

3. Sanford S. Miller, Petrue T. Mocanu and Maxwell O. Reade, *All α-convex functions are starlike*, Rev. Roumaine Math. Pures Appl. **17** (1972), 1395–1397.

4. ______, *All α-convex functions are univalent and starlike*, Proc. Amer. Math. Soc. **37** (1973), 553–554.

5. V. A. Zmorovich and I. K. Korobkova, *On the order of starlikeness for the class of α-convex functions for $\alpha \leqslant -2$*, Mathematical Analysis and Probability Theory (V. S. Korolyuk, editor), "Naukova Dumka", Kiev, 1978, pp. 63–66. (Russian)

6. V. A. Zmorovich and V. A. Pokhilevich, *On α-convex functions for $\alpha < 0$*, Ukrain. Mat. Zh. **31** (1979), 431–433; English transl. in Ukrainian Math. J. **31** (1979).

7. P. Eenigenburg and D. Nelson, *On the order of starlikeness for alpha-convex functions*, Mathematica (Cluj) **18 (41)** (1976), 143–146.

8. V. A. Zmorovich and I. K. Korobkova, *On the order of starlikeness of certain subclasses of α-convex functions for $\alpha > 0$*, Ukrain. Mat. Zh. **30** (1978), 536–539; English transl. in Ukrainian Math. J. **30** (1978).

9. Stephan Ruscheweyh and Vikramaditya Singh, *On certain extremal problems for functions with positive real part*, Proc. Amer. Math. Soc. **61** (1976), 329–334.

Translated by R. P. BOAS

Amer. Math. Soc. Transl.
(2) Vol. **127**, 1986

On Representation of Entire Functions Positive on the Real Axis, or on a Semiaxis, or Outside a Finite Interval*

M. G. KREĬN AND A. A. NUDEL′MAN

Abstract. This article deals with the generalization of the following well-known propositions to arbitrary entire functions.

PROPOSITION I. *Every algebraic polynomial $f(z)$ positive on $(-\infty, \infty)$ admits a representation*

$$f(z) = A^2(z) + B^2(z), \tag{1}$$

*where $A(z)$ and $B(z)$ are real algebraic polynomials which can be chosen so that all their roots are simple and real, and alternate.***

PROPOSITION II. *Every algebraic polynomial $f(z)$ positive on $[0, \infty)$ admits a representation*

$$f(z) = C^2(z) + zD^2(z), \tag{2}$$

where $C(z)$ and $D(z)$ are real algebraic polynomials. These polynomials can be chosen (in a unique way) so that the following conditions hold:

1) *The degree of $D(z)$ is either equal to that of $C(z)$ or less than it by* 1.

2) $C(0) > 0$ *and* $D(0) > 0$.

3) *The roots $\xi_1, \xi_2, \ldots$ and $\eta_1, \eta_2, \ldots$ of the respective polynomials $C(z)$ and $D(z)$ are all simple and positive, and alternate in such a way that $0 < \xi_1 < \eta_1 < \xi_2 < \eta_2 < \cdots$.*

If polynomials $A(z)$ and $B(z)$, whose roots are real and simple and alternate, are suitably normalized, then their ratio can be represented in the form

$$\frac{B(z)}{A(z)} = \alpha + \beta z + \sum_{j=1}^{m} \frac{\rho_j}{\xi_j - z},$$

where α is a real number, $\beta \geqslant 0$, $\rho_j > 0$, and the ξ_j are the roots of $A(z)$.

1980 *Mathematics Subject Classification*. Primary 30C15, 30D10; Secondary 30C10, 30D05, 30D15.

*Translation of Mat. Issled. Vyp. 61 (1981), 40–59; MR **82g**: 30050.

***Editor's note*. The term "alternate" here means that between any two consecutive roots of one polynomial there is a root of the other polynomial.

The ratio B/A is thus in the class $\mathscr{R}$ of functions holomorphic in the open upper half-plane and having nonnegative imaginary part there.

Therefore, in generalizing Proposition I we shall require that the ratio B/A be in the class $\mathscr{R}$.

Starting from analogous considerations, in generalizing Proposition II we shall require that the ratio D/C belong to the class $\mathscr{S}$ of functions in $\mathscr{R}$ that are regular and nonnegative on the negative semiaxis.

Closely related problems have been solved by a number of authors under various assumptions on the growth of an entire function $f(z)$ ([1]–[4]; see also [5]). In the generalization of Proposition I the problem was posed of representing a function $f(z)$ nonnegative on $(-\infty, \infty)$ in the form

$$f(z) = h(z)\bar{h}(z), \tag{3}$$

where $h(z)$ is an entire function growing "twice as slowly" as $f(z)$ and not having roots in the lower half-plane. For real $A(z)$ and $B(z)$ the condition $B/A \in \mathscr{R}$ implies that $\operatorname{Im}(B/A) < 0$ when $\operatorname{Im} z < 0$, and this implies that the equation $B/A = i$, and with it the function $h(z) = A(z) + iB(z)$, does not have roots in the lower half-plane, so that under this condition the representation (1) simultaneously gives us the factorization (3). The converse, however, is false: the fact that $h(z) = A(z) + iB(z)$ does not have roots in the lower half-plane does not in general imply that B/A (or $-B/A$) $\in \mathscr{R}$ [3].

Thus, here Propositions I and II can be generalized without any restrictions on the growth of $f(z)$ under more stringent connections between the functions $A(z)$ and $B(z)$ (or $C(z)$ and $D(z)$) (Theorems 1 and 3). These connections are equivalent to the requirement that $|h(z)/\bar{h}(z)| \leqslant 1$ when $\operatorname{Im} z \geqslant 0$. However, to establish Theorem 1 we needed to add little to what was already known. In particular, the sufficiency of the condition (9) in Theorem 1 was obtained by another path in the book [6] of de Branges. In addition to Theorems 1 and 3 we establish formulas enabling us to find all possible pairs $A(z)$, $B(z)$ and $C(z)$, $D(z)$ of functions subject to the indicated connections and giving the respective representations (1) and (3) (Theorems 2 and 4). The need for these formulas arose in solving an inverse problem in the theory of the spectral functions of a string [7].

Proposition II for a $(2n)$th degree polynomial $f(z)$ positive on $(-\infty, a\,]$ and $[\,b, \infty)$ leads to the following result via the transformation

$$f(z) \to p(\zeta) = (1-\zeta)^{2n} f((b\zeta - a)/(1-\zeta)).$$

PROPOSITION III. *Every polynomial $f(z)$ positive on $(-\infty, a\,]$ and $[\,b, \infty)$ admits the representation*

$$f(z) = A^2(z) + (z-a)(z-b)B^2(z), \tag{4}$$

where the polynomials $A(z)$ and $B(z)$ can be chosen (in a unique way) so that $A(b) > 0$ and the ratio $(z-a)B(z)/A(z)$ is in the class $\mathscr{S}(E)$ of functions in $\mathscr{R}$ that are regular and nonnegative in (a, b).

This proposition also generalizes to the case of entire functions (Theorem 5), but for obvious reasons the generalization cannot be obtained from the generalization of Proposition II by the aforementioned transformation for polynomials.

A representation of type (4) was obtained by Rybalko [8] for entire functions of exponential type under somewhat different conditions on $A(z)$ and $B(z)$.

Representations (1), (2), and (4) for polynomials played an important role in the investigations of Tchebycheff and his school in finding solutions of a number of extremal problems and then later in other problems in the constructive theory of functions.

In solving the problem of the determination of all the pairs (A, B) satisfying the required conditions we had to consider Pell's functional equation

$$X^2(z) + R(z)Y^2(z) = 1$$

with $R(z) = (z-a)(z-b)$. The study of this equation in the case where R is a polynomial and X and Y must be polynomials goes back to the memoir [9] of Abel. Some results relating to this equation for the case where $R(z)$ is an entire function of this or that class and $X(z)$ and $Y(z)$ are entire functions of no greater than exponential type have been obtained in the theory of λ-zones of stability for solutions of the Sturm-Liouville problem with periodic coefficients [10], [11].

It was observed fairly recently that the representations (1), (2), and (4) play a fundamental role in the construction of natural continuous orthogonal systems of functions in the space $L^2(-\infty, \infty; \Omega)$, where Ω is a weight of a special form (see [12]–[16]). The authors hope to cover a new approach to these questions in a separate paper.

§1. Auxiliary propositions

1. We give some facts needed below about functions in the classes $\mathscr{R}$, $\mathscr{S}(E)$, and $\mathscr{G}$ (for a detailed exposition see [17] and [18]).

As already noted, a function $F(z)$ belongs to the class $\mathscr{R}$ if it is holomorphic in the upper half-plane and $\operatorname{Im} F(z) \geqslant 0$ when $\operatorname{Im} z > 0$.([1])

It is well known that $F \in \mathscr{R}$ if and only if the representation

$$F(z) = \alpha + \beta z + \int_{-\infty}^{\infty} \left(\frac{1}{t-z} - \frac{t}{1+t^2} \right) d\sigma(t) \qquad (\operatorname{Im} z > 0), \tag{5}$$

is valid, where $\bar{\alpha} = \alpha$, $\beta \geqslant 0$, $d\sigma(t) \geqslant 0$, and $\int_{-\infty}^{\infty} (1+t^2)^{-1}\, d\sigma(t) < \infty$. From the Stieltjes inversion formula it follows that if a function $F \in \mathscr{R}$ is regular and real in some interval of the real axis, then $d\sigma(t) = 0$ in this interval. It is easy to see that $F \in \mathscr{R}$ if and only if $-1/F \in \mathscr{R}$.

For a finite or infinite interval $[a, b]$ let $E = (-\infty, \infty) \setminus [a, b]$ and denote by $\mathscr{S}(E)$ the class of functions in $\mathscr{R}$ that are regular and nonnegative in (a, b). If $a = -\infty$ and $b = 0$, then the corresponding class of functions is denoted by $\mathscr{S}$. A

([1]) The requirement that the function be holomorphic can be replaced by the requirement that it be meromorphic in the upper half-plane; if this holds, then the condition that $\operatorname{Im} F(z) \geqslant 0$ when $\operatorname{Im} z > 0$ implies that $F(z)$ is holomorphic for $\operatorname{Im} z > 0$.

function $F(z)$ belongs to $\mathscr{S}$ if and only if $-1/zF(z) \in \mathscr{S}$, and also if and only if $zF(z^2) \in \mathscr{R}$.(²) In connection with the last assertion we note that a function $G(z)$ admits a representation $G(z) = zF(z^2)$ with $F \in \mathscr{S}$ if and only if it belongs to the class $\mathscr{G}$ of functions in $\mathscr{R}$ with the property $G(-\bar{z}) = \overline{G(z)}$.

If (a, b) is a finite interval and if we set $\zeta = (b - z)/(a - z)$ and $\Phi(\zeta) = F(z)$, then $F \in \mathscr{S}(E)$ if and only if $\Phi \in \mathscr{S}$. The use of the criteria given above for Φ to belong to $\mathscr{S}$ gives us that $F \in \mathscr{S}(E)$ if and only if $(z - a)/(b - z)F(z) \in \mathscr{S}(E)$, and also if and only if $\zeta F[(b - a\zeta^2)/(1 - \zeta^2)] \in \mathscr{R}$, i.e., if and only if

$$\frac{b - z}{a - z} F\left(\frac{ab - z^2}{a + b - 2z}\right) \in \mathscr{R}.$$

This criterion is used with $a = -1$ and $b = 1$ in what follows; in this case the last condition takes the form

$$\frac{z - 1}{z + 1} F\left(\frac{z^2 + 1}{2z}\right) \in \mathscr{R}.$$

It is convenient to include the function $F(z) \equiv \infty$ in each of the classes $\mathscr{R}, \mathscr{G}$, and $\mathscr{S}(E)$.

2. We mention propositions that can be regarded as "laws of composition" for the functions in this or that class that do not lead out of the corresponding class.

a) *If* $u, v \in \mathscr{R}$, *then*

$$F = \frac{u + v}{1 - uv} \in \mathscr{R}.$$

Indeed, $1 - u(z)v(z) \neq 0$ when $\operatorname{Im} z > 0$, since $\operatorname{Im} u(z) > 0$ and $\operatorname{Im}(1/v(z)) < 0$, and

$$\operatorname{Im} F(z) = \frac{\operatorname{Im} u(z)\left(1 + |v(z)|^2\right) + \operatorname{Im} v(z)\left(1 + |u(z)|^2\right)}{|1 - u(z)v(z)|^2}.$$

b) *If* $u, v \in \mathscr{S}$, *then*

$$F(z) = \frac{u(x) + v(z)}{1 - zu(z)v(z)} \in \mathscr{S}.$$

Indeed, setting $U(z) = zu(z^2)$ $(\in \mathscr{R})$ and $V(z) = zv(z^2)$ $(\in \mathscr{R})$, we get that

$$zF(z^2) = \frac{U(z) + V(z)}{1 - U(z)V(z)} \in \mathscr{R}.$$

For simplicity we assume that $a = -1$ and $b = 1$.

c) *If* $u, v \in \mathscr{S}(E)$, *then*

$$F(z) = \frac{u(z) + v(z)}{1 - [(z - 1)/(z + 1)]\, u(z)v(z)} \in \mathscr{S}(E).$$

(²) It is natural to regard a function $F(z) \in \mathscr{S}(E)$ as being analytically continued into the lower half-plane by the formula $F(\bar{z}) = \overline{F(z)}$.

Indeed, if

$$U(z)=\frac{z-1}{z+1}u\left(\frac{z^2+1}{2z}\right)\quad(\in\mathscr{R}),\qquad V(z)=\frac{z-1}{z+1}v\left(\frac{z^2+1}{2z}\right)\quad(\in\mathscr{R}),$$

then

$$\frac{z-1}{z+1}F\left(\frac{z^2+1}{2z}\right)=\frac{U(z)+V(z)}{1-U(z)V(z)}\in\mathscr{R}.$$

§2. Some special problems of the Nevanlinna-Pick type

1. PROBLEM $(\mathscr{R}; i)$. Let $Z=\{\zeta_k\}$ ($\operatorname{Im}\zeta_k>0$, $\zeta_k\to 1$) be a given sequence of complex numbers. Denote by $(\mathscr{R}; i)$ the problem of finding all the functions $F\in\mathscr{R}$ such that Z is the sequence of roots of the difference $F-i$ (and of its corresponding derivatives if several of the values ζ_k coincide). One such function is $F_0(\zeta)\equiv i$. For the existence of other solutions of Problem $(\mathscr{R}; i)$ it is necessary that the Blaschke product

$$\varphi(\zeta)=\prod_k\frac{\zeta-\zeta_k}{\zeta-\bar\zeta_k}$$

converge, and this is equivalent to the condition

$$\sum_k\operatorname{Im}\zeta_k<\infty. \tag{6}$$

We show that this condition is also sufficient for the existence of solutions different from F_0, and we find a formula giving all the solutions.

Let F be some solution of Problem $(\mathscr{R}; i)$. Since $\operatorname{Im}F(\zeta)\geqslant 0$ when $\operatorname{Im}\zeta>0$, the function $(F(\zeta)-i)/(F(\zeta)+i)$ does not exceed in modulus when $\operatorname{Im}\zeta>0$ and remains so after division by the Blaschke product $\varphi(\zeta)$, because the set of roots of φ (counting multiplicities) is in the set of roots of $F-i$, and $|\varphi(\zeta)|=1$ when $\operatorname{Im}\zeta=0$. Thus,

$$\frac{F(\zeta)-i}{F(\zeta)+i}=\varphi(\zeta)\frac{\Omega(\zeta)-i}{\Omega(\zeta)+i}, \tag{7}$$

where $\Omega\in\mathscr{R}$. Conversely, formula (7) assigns to each $\Omega\in\mathscr{R}$ some solution of Problem $(\mathscr{R}; i)$. Consequently, the set of all solutions of $(\mathscr{R}; i)$ can be described by the formula

$$F=\frac{[(\varphi+1)/2]\Omega+(\varphi-1)/2i}{-[(\varphi-1)/2i]\Omega+(\varphi+1)/2},$$

where Ω runs through the class $\mathscr{R}$.

2. PROBLEM $(\mathscr{S}; i/\sqrt{\zeta})$.(3) Let $Z=\{\zeta_k\}$ be a numerical sequence converging to 1 and located in the upper half-plane and, possibly, on the negative real semiaxis. Here the number of negative terms of the sequence is assumed to be even. Problem $(\mathscr{S}; i/\sqrt{\zeta})$ is understood to be that of finding all $F\in\mathscr{S}$ such that Z is the sequence of roots (counting multiplicities) of the difference $F(\zeta)-i\sqrt{\zeta}$.

---

(3) Here and below we have in mind the branch of $\sqrt{\zeta}$ with imaginary part positive in the upper half-plane, defined in the plane cut along $[0,\infty)$.

One of its solutions is the function $F_0(\zeta) = i/\sqrt{\zeta}$. For the existence of other solutions condition (6) is necessary, and it also turns out to be sufficient. To prove this and to find all the solutions we consider the auxiliary Problem $(\mathscr{R}; i)$ associated with the sequence composed of the numbers $\sqrt{\zeta_k}$ and $-\sqrt{\bar{\zeta}_k}$ when $\operatorname{Im}\zeta_k > 0$ and of $\sqrt{\zeta_k}$ when $\zeta_k < 0$. Condition (6) implies the convergence as $\zeta_k \to 1$ of the series $\sum \operatorname{Im}\sqrt{\zeta_k}$ and, consequently, of the product

$$\psi(\zeta) = \prod_{\operatorname{Im}\zeta_k > 0} \frac{\zeta - \sqrt{\zeta_k}}{\zeta - \sqrt{\bar{\zeta}_k}} \prod_{\zeta_k < 0} \left(\zeta - \sqrt{\zeta_k}\right).$$

The Blaschke product $\varphi(\zeta)$ corresponding to the auxiliary problem can be represented in the form $\varphi(\zeta) = \psi(\zeta)/\psi(-\zeta)$, so that the solutions of this problem can be described by the formula

$$G(\zeta) = \frac{[(\psi(\zeta) + \psi(-\zeta))/2]\Omega(\zeta) + (\psi(\zeta) - \psi(-\zeta))/2i}{-[(\psi(\zeta) - \psi(-\zeta))/2i]\Omega(\zeta) + (\psi(\zeta) + \psi(-\zeta))/2},$$

where $\Omega \in \mathscr{R}$. It is easy to see that $G \in \mathscr{G}$ if $\Omega \in \mathscr{G}$ and, conversely, $\Omega \in \mathscr{G}$ if $G \in \mathscr{G}$. Setting $\Omega(\zeta) = \zeta\omega(\zeta^2)$ $(\omega \in \mathscr{S})$ and $G(\zeta) = \zeta F(\zeta^2)$ $(F \in \mathscr{S})$ and replacing ζ by $\sqrt{\zeta}$, we get that the solutions of Problem $(\mathscr{S}; i\sqrt{\zeta})$ are described by the formula

$$F(\zeta) = \frac{\left[(\psi(\sqrt{\zeta}) + \psi(-\sqrt{\zeta}))/2\right]\omega(\zeta) + \left(\psi(\sqrt{\zeta}) - \psi(-\sqrt{\zeta})\right)/2i\sqrt{\zeta}}{-\zeta\left[(\psi(\sqrt{\zeta}) - \psi(-\sqrt{\zeta}))/2i\sqrt{\zeta}\right]\omega(\zeta) + \left(\psi(\sqrt{\zeta}) + \psi(-\sqrt{\zeta})\right)/2}, \quad (8)$$

where ω runs through the class $\mathscr{S}$.

3. Problem $(\mathscr{S}(E); \sqrt{(z-a)/(b-z)})$. Let $Z = \{z_k\}$ $(z_k \to \infty)$ be a sequence of complex numbers located in the upper half-plane and, possibly, in (a, b); the number of terms in (a, b) is assumed to be even. It is required to describe all the $\Phi \in \mathscr{S}(E)$ for which Z is the sequence of roots (counting multiplicities) of the difference $\varphi(z) - \sqrt{(z-a)/(b-z)}$. The substitution $\zeta = (z-b)/(z-a)$ reduces this problem at once to Problem $(\mathscr{S}; i/\sqrt{\zeta})$ for the sequence $\{\zeta_k\}$, $\zeta_k = (z_k - b)/(z_k - a)$. Moreover, condition (6) in this case is equivalent to

$$\sum_k \operatorname{Im}(-1/z_k) < \infty.$$

Under this condition a description of all the solutions Φ of Problem $(\mathscr{S}(E); \sqrt{(z-a)/(b-z)})$ is given with the help of formula (8), with $\zeta = (b-z)/(z-a)$ and $\varphi(z) = F(z)$.

§3. Entire functions positive in $(-\infty, \infty)$

The following is used below.

Assertion A) [19]. *Suppose that $\Phi(z)$ is a meromorphic function with $|\Phi(z)| \leqslant 1$ when $\operatorname{Im} z > 0$ and $|\Phi(z)| \geqslant 1$ when $\operatorname{Im} z \leqslant 0$. Then Φ admits a representation*

$$\varphi(z) = e^{(a+bz)i}\chi(z),$$

where $\bar{a} = a$, $b \geqslant 0$, *and*

$$\chi(z) = \prod_j \left[\left(1 - \frac{z}{z_j}\right) \Big/ \left(1 - \frac{z}{\bar{z}_j}\right) \right]$$

with $\operatorname{Im} z_j > 0$ (*if* Φ *does not have roots, then* $\chi(z) \equiv 1$). *Moreover, if there are infinitely many roots* z_j, *then*

$$\sum_j \operatorname{Im}(-1/z_j) < \infty, \tag{9}$$

which ensures convergence of the product χ.

THEOREM 1. *For an entire function* $f(z)$ *positive on* $(-\infty, \infty)$ *to admit a representation*

$$f(z) = A^2(z) + B^2(z), \tag{10}$$

where A *and* B *are real entire functions not having common roots and such that* $B/A \in \mathcal{R}$, *it is necessary and sufficient that* (9) *hold for the roots* z_j *of* $f(z)$ *lying in the upper half-plane.*

PROOF. The necessity follows from the facts that for $B/A \in \mathcal{R}$ the function $\varphi = (A + iB)/(A - iB)$ satisfies the conditions of Assertion A), and its roots coincide with those of f lying in the upper half-plane. Let us prove the sufficiency.

Using the canonical Weierstrass product, we represent f in the form

$$f(z) = e^{g(z)} \prod_j \left(1 - \frac{z}{z_j}\right) e^{p_j(z)} \prod_j \left(1 - \frac{z}{\bar{z}_j}\right) e^{\bar{p}_j(z)} = e^{g(z)} \prod_j \left(1 - \frac{z}{z_j}\right)\left(1 - \frac{z}{\bar{z}_j}\right) e^{2r_j(z)}.$$

Here g is a real entire function, and the $2r_j(z) = p_j(z) + \bar{p}_j(z)$ are real polynomials.

Since $f(z) = e^{g(z)} P^2(z) \chi(z)$, where $P(z) = \prod_j (1 - (z/z_j)) e^{r_j(z)}$, introduction of the functions

$$A_0(z) = e^{g(z)/2} P(z)(\chi(z) + 1)/2,$$
$$B_0(z) = e^{g(z)/2} P(z)(\chi(z) - 1)/2i$$

leads to the equality

$$f(z) = A_0^2(z) + B_0^2(z).$$

We note the following properties of A_0 and B_0:

a) A_0 and B_0 are entire functions (since each pole of the meromorphic function $\chi \pm 1$ is a root of P of the same multiplicity).

b) A_0 and B_0 are real functions (since $\bar{\chi}(z) = 1/\chi(z)$ and $\bar{P}(z)/\chi(z) = P(z)$).

c) A_0 and B_0 do not have common roots (this is obvious).

d) $B_0/A_0 = i(1 - \chi)/(1 + \chi) \in \mathcal{R}$ (since $|\chi(z)| < 1$ when $\operatorname{Im} z > 0$).

This and property c) give us, in particular, that all the roots of A_0 and B_0 are real and simple, and that they alternate.

A pair (A, B) of real entire functions without common roots that realizes the representation (10) with the condition $B/A \in \mathscr{R}$ satisfied is called an *$\mathscr{R}$-pair* for f.

THEOREM 2. *The set of all $\mathscr{R}$-pairs for f is obtained from the pair (A_0, B_0) by the formulas*

$$\begin{aligned} A(z) &= A_0(z)\cos(\alpha + \beta z) - B_0(z)\sin(\alpha + \beta z), \\ B(z) &= A_0(z)\sin(\alpha + \beta z) + B_0(z)\cos(\alpha + \beta z), \end{aligned} \tag{11}$$

where α runs through the set of real numbers and β through the set of nonnegative numbers.

PROOF. For arbitrary $\alpha = \bar{\alpha}$ and $\beta \geqslant 0$ it is obvious that A and B are real entire functions without common roots, and $A^2 + B^2 = A_0^2 + B_0^2 = f$. Since $u = \tan(\alpha + \beta z) \in \mathscr{R}$ for $\alpha = \bar{\alpha}$ and $\beta \geqslant 0$, and $v = B_0/A_0 \in \mathscr{R}$, it follows that $B/A = (u + v)/(1 - uv) \in \mathscr{R}$, i.e., (A, B) is an $\mathscr{R}$-pair for f.

Conversely, let (A, B) be an $\mathscr{R}$-pair for f. Then, by Assertion A) and the fact that $\chi = (A_0 + iB_0)/(A_0 - iB_0)$, for some $\alpha = \bar{\alpha}$ and $\beta \geqslant 0$ we have

$$\frac{A(z) + iB(z)}{A(z) - iB(z)} = e^{2(\alpha + \beta z)i} \frac{A_0(z) + iB_0(z)}{A_0(z) - iB_0(z)}. \tag{12}$$

After multiplying by the denominators and separating the real and imaginary parts for real z, we find by simple transformations that the identity

$$\begin{aligned} &(A_0(z)A(z) + B_0(z)B(z))\sin(\alpha + \beta z) \\ &\quad = (A_0(z)B(z) - A(z)B_0(z))\cos(\alpha + \beta z) \end{aligned}$$

holds for real z, and hence for all z; this can be rewritten in the form $AB_1 = BA_1$, where

$$\begin{aligned} A_1(z) &= A_0(z)\cos(\alpha + \beta z) - B_0(z)\sin(\alpha + \beta z), \\ B_1(z) &= A_0(z)\sin(\alpha + \beta z) + B_0(z)\cos(\alpha + \beta z). \end{aligned} \tag{13}$$

Since $f = A_1^2 + B_1^2 = A_1^2 + (A_1B/A)^2 = (A_1/A)^2 f$, it follows that $A_1^2(z) = A^2(z)$ and $B_1^2(z) = B^2(z)$. Therefore, $A(z) = \pm A_1(z)$ and $B(z) = \pm B_1(z)$, and the signs must be the same, since $B/A \in \mathscr{R}$ and $B_1/A_1 \in \mathscr{R}$.

If $A = A_1$ and $B = B_1$, then (13) coincides with (11); but if $A = -A_1$ and $B = -B_1$, then (13) passes into (11) as a result of replacing α by $\alpha + \pi$ in (13). This proves the theorem.

REMARK. As already remarked in the Introduction, if we associate the function $h = A + iB$ with an arbitrary $\mathscr{R}$-pair (A, B) for f, then we get an entire function that realizes the representation $f(z) = h(z)\bar{h}(z)$ and has no roots in the lower half-plane. By (11),

$$h^2(z) = f(z)h(z)/\bar{h}(z) = f(z)\chi(z)e^{2(\alpha + \beta z)i},$$

where $\alpha = 0$ and $\beta = 0$ for $h_0 = A_0 + iB_0$, so repetition of arguments in [5, Appendix V] in our case leads to the following conclusions:

a) If f is an entire function of *proximate* order $\rho(r)$ and type σ, then h_0 has the same order and type $\sigma/2$.

b) If $\rho > 1$ here, then for any $\mathscr{R}$-pair the function h has order $\rho(r)$ and type $\sigma/2$.

§4. Entire functions positive on $[0, \infty)$

Passing to entire function positive on $[0, \infty)$, we define an *$\mathscr{S}$-pair* for f to be a pair (C, D) of real entire functions that have no common roots, realize the representation

$$f(z) = C^2(z) + zD^2(z), \tag{14}$$

and satisfy the conditions

$$C(0) > 0, \qquad D/C \in \mathscr{S}. \tag{15}$$

THEOREM 3. *For an entire function f positive on $[0, \infty)$ there exists an $\mathscr{S}$-pair (C, D) if and only if the roots z_j of f located in the closed upper half-plane satisfy the condition*

$$\sum_j \operatorname{Im}\left(-1/\sqrt{z_j}\right) < \infty. \tag{16}$$

PROOF. If (14) and (15) hold for f, then by setting $G(\zeta) = f(\zeta^2)$, $A(\zeta) = C(\zeta^2)$, and $B(\zeta) = \zeta D(\zeta^2)$, we get that (A, B) is an $\mathscr{R}$-pair for the entire function G, which is positive on $(-\infty, \infty)$. The facts that A and B are real and have no common roots are obvious, and the relation $B/A \in \mathscr{R}$ follows from the fact that $D/C \in \mathscr{S}$. Since the roots of G in the upper half-plane are the numbers $\sqrt{z_j}$ and $-\sqrt{\bar{z}_j}$, we conclude from Theorem 1 that condition (16) holds.

Conversely, if (16) holds, then it is possible to construct an $\mathscr{R}$-pair (A_0, B_0) for the function $G(\zeta) = f(\zeta^2)$. Since the roots of G are symmetric with respect to the imaginary axis, the g in the representation $G(\zeta) = e^{g(\zeta)}P^2(\zeta)\chi(\zeta)$ is an even function, and $P(\zeta)$ and $\chi(\zeta)$ have the additional properties $P(-\zeta) = \bar{P}(\zeta)$ and $\chi(-\zeta) = \bar{\chi}(\zeta)$. Therefore, A_0 is even and B_0 is odd, so that the pair (C_0, D_0) with $A_0(\zeta) = C_0(\zeta^2)$ and $B_0(\zeta) = \zeta D_0(\zeta^2)$ is an $\mathscr{S}$-pair for f.

THEOREM 4. 1) *The set of all $\mathscr{S}$-pairs (C, D) for f is obtained from the "main pair" (C_0, D_0) by the formulas*

$$\begin{aligned} C(z) &= C_0(z)\cos\beta\sqrt{z} - \sqrt{z}\, D_0(z)\sin\beta\sqrt{z}, \\ D(z) &= C_0(z)(\sin\beta\sqrt{z})/\sqrt{z} + D_0(z)\cos\beta\sqrt{z}, \end{aligned} \tag{17}$$

where β runs through the set of nonnegative numbers.

2) *The main pair (C_0, D_0) has the following extremal property: for any other $\mathscr{S}$-pair (C, D) for f and for each $x > 0$*

$$\operatorname{sgn}\left(\frac{D(-x)}{C(-x)} - \frac{D_0(-x)}{C_0(-x)}\right) = \operatorname{sgn} f(-x).$$

PROOF. 1) Each $\mathscr{S}$-pair (C, D) for f generates an $\mathscr{R}$-pair (A, B) for the function $G(\zeta) = f(\zeta^2)$: $A(\zeta) = C(\zeta^2)$, $B(\zeta) = \zeta D(\zeta^2)$, in which $A(\zeta)$ is even, $B(\zeta)$ is odd, and $A(0) > 0$. Conversely, each $\mathscr{R}$-pair (A, B) for the function $G(\zeta) = f(\zeta^2)$, in which $A(\zeta)$ is even, $B(\zeta)$ is odd, and $A(0) > 0$ generates an $\mathscr{S}$-pair $(C(z), D(z))$ for $f(z)$: $C(z) = A(\sqrt{z})$, $D(z) = B(\sqrt{z})/\sqrt{z}$. This property is enjoyed by precisely those pairs (A, B) obtained by (11) with $\alpha \equiv 0 \pmod{2\pi}$ and with $\beta \geqslant 0$, and this gives us the first assertion of the theorem.

2) According to (16),

$$\frac{D(z)}{C(z)} = \frac{C_0(z)\omega(z) + D_0(z)}{-zD_0(z)\omega(z) + C_0(z)},$$

where $\omega(z) = (\tan \beta\sqrt{z})/\sqrt{z} \in \mathscr{S}$ $(\beta > 0)$. Therefore,

$$\frac{D(z)}{C(z)} - \frac{D_0(z)}{C_0(z)} = \frac{f(z)\omega(z)}{C_0(z)(-zD_0(z)\omega(z) + C_0(z))}. \tag{18}$$

Since $D_0/C_0 \in \mathscr{S}$, this function is positive for $z = -x < 0$, and since $C_0(z)$ does not have zeros on $(-\infty, 0]$ and $C_0(0) > 0$, it follows that $C_0(-x) > 0$ and $D_0(-x) > 0$. Moreover, $\omega(-x) > 0$. Therefore, the sign of the difference (18) coincides with the sign of $f(-x)$.

REMARK. If, in addition to (16), we have the condition $\Sigma_j(1/|z_j|) < \infty$, so that the product

$$Q(z) = \prod_{\operatorname{Im} z_j > 0} \left[\left(\frac{1 - z}{\sqrt{z_j}}\right)\left(\frac{1 + z}{\sqrt{\bar{z}_j}}\right)\right] \prod_{z_j < 0} \left(\frac{1 - z}{\sqrt{z_j}}\right)$$

converges, and f itself admits the representation

$$f(z) = e^{g(z)} \prod_{\operatorname{Im} z_j > 0} \left[\left(\frac{1 - z}{z_j}\right)\left(\frac{1 - z}{\bar{z}_j}\right)\right] \prod_{z_j < 0} \left(\frac{1 - z}{z_j}\right),$$

then C_0 and D_0 can be defined by the formulas

$$C_0(z) = e^{g(z)/2}(Q(\sqrt{z}) + Q(-\sqrt{z}))/2,$$
$$D_0(z) = e^{g(z)/2}(Q(\sqrt{z}) - Q(-\sqrt{z}))/2i\sqrt{z}.$$

Moreover, since all the roots are positive for each of a pair of functions C and D forming an $\mathscr{S}$-pair for f, the Lindelöf theorem [5, Chapter I, §11, Theorem 15] leads to the conclusion that $e^{-g(z)/2}C(z)$ and $e^{-g(z)/2}D(z)$ have genus zero.

§5. Entire functions positive on E

THEOREM 5. *For an entire function $f(z)$ positive for $z \in E = (-\infty, \infty)\setminus(a, b)$ to admit a representation*

$$f(z) = A^2(z) + (z - a)(z - b)B^2(z), \tag{19}$$

where A and B are real entire functions without common roots and $(z - a)B(z)/A(z) \in \mathscr{S}(E)$, it is necessary and sufficient that the roots z_j of f in the upper half-plane satisfy the condition

$$\sum_j \operatorname{Im}(-1/z_j) < \infty. \tag{20}$$

PROOF. *Necessity*. From (19),

$$\left(\sqrt{\frac{z-a}{b-z}}-\frac{(z-a)B(z)}{A(z)}\right)\left(\sqrt{\frac{z-a}{b-z}}+\frac{(z-a)B(z)}{A(z)}\right)=\frac{z-a}{(b-z)A^2(z)}.$$

The second factor on the left-hand side is positive for $z \in (a, b)$ and has a positive imaginary part in the upper half-plane. Consequently, the roots of the first factor other than $z = a$ that lie in (a, b) or in the upper half-plane coincide (counting multiplicities) with the corresponding roots of f. Therefore, the function $F(z) = (z - a)B(z)/A(z)$ is a solution, different from

$$F_0(z) = \sqrt{(z-a)/(b-z)}\,,$$

of Problem $(\mathscr{S}(E); \sqrt{(z-a)/(b-z)})$ associated with the sequence $\{z_j\}$ of roots of f; hence (20) holds (see §1).

With what follows in mind, it is appropriate to observe that, along with the function $F(z) = (z - a)B(z)/A(z)$, the function

$$(z-a)/(b-z)F(z) = A(z)/(b-z)B(z)$$

also belongs to the class $\mathscr{S}(E)$ and, as is easy to see, is also a solution of Problem $(\mathscr{S}(E); \sqrt{(z-a)/(b-z)})$.

Sufficiency. The decomposition of f into a canonical Weierstrass product leads to the representation

$$f(z) = e^{g(z)} \prod_{\operatorname{Im} z_j>0} \left[\left(1-\frac{z}{z_j}\right)\left(1-\frac{z}{\bar z_j}\right)\right] e^{2r_j(z)} \prod_{k=1}^{2\nu} (z-x_k),$$

where g is a real entire function, $r_j(z)$ is a real polynomial, and the x_k are the roots of f located in (a, b). For simplicity it is assumed that $a = -1$ and $b = 1$. If condition (20) holds, then the Blaschke product

$$x(z) = \prod_j \left[\left(1-\frac{z}{z_j}\right)\Big/\left(1-\frac{z}{\bar z_j}\right)\right]$$

converges, so that

$$f(z) = e^{g(z)}P^2(z)\chi(z)Q(z),$$

where $P(z) = \prod_j(1 - z/\bar z_j)e^{r_j(z)}$ and $Q(z) = \prod_{k=1}^{2\nu}(z - x_k)$. Now let $\zeta = (z - 1)/(z + 1)$, $\zeta_j = (z_j - 1)/(z_j + 1)$, and $\varepsilon_k = (x_k - 1)/(x_k + 1)$. With the notation of §2,

$$\chi(z) = \prod_j\left[(\zeta-\zeta_j)/(\zeta-\bar\zeta_j)\right]\prod_j\left[(1+\bar\zeta_j)/(1+\zeta_j)\right] = \varphi(\zeta)/\varphi(-1),$$

and the product $\chi(z)Q(z)$ can be represented in the form

$$\chi(z)Q(z) = \frac{\psi(\sqrt{\zeta})\psi(-\sqrt{\zeta})}{\varphi(-1)(1-\zeta)^{2\nu}}. \tag{21}$$

We introduce the functions

$$A_0(z) = e^{g(z)/2}P(z)\frac{\psi(\sqrt{\zeta}) + \psi(-\sqrt{\zeta})}{2\sqrt{\varphi(-1)}\,(1-\zeta)^{\nu}} = e^{g(z)/2}P(z)a_0(z),$$
$$B_0(z) = e^{g(z)/2}P(z)\frac{\psi(\zeta) - \psi(-\sqrt{\zeta})}{4i\sqrt{\varphi(-1)}\,(1-\zeta)^{\nu-1}\sqrt{\zeta}} = e^{g(z)/2}P(z)b_0(z). \tag{22}$$

It is easy to see that

$$A_0^2(z) + (z^2-1)B_0^2(z) = A_0^2(z) + \frac{4\zeta}{(1-\zeta)^2}B_0^2(z) = f(z).$$

We determine the properties of the functions A_0 and B_0.

a) The functions $a_0(z)$ and $b_0(z)$ are univalent. Since the singularities of $\psi(\zeta)$ are the poles $\sqrt{\zeta_j}$ of order ν_j and they accumulate at $\zeta = 1$, the singularities of $a_0(z)$ and $b_0(z)$ are the poles $\bar{z}_j$ of order ν_j and they accumulate at the point $z = \infty$, i.e., a_0 and b_0 are meromorphic functions whose poles coincide (counting multiplicities) with the roots of P. Thus, A_0 and B_0 are entire functions. It is clear that they do not have common zeros.

b) Let us verify that A_0 and B_0 are real functions.

Indeed, recalling that

$$\psi(\zeta) = \prod_{\operatorname{Im}\zeta_j>0}\left[\frac{\left(\zeta - \sqrt{\zeta_j}\right)}{\left(\zeta - \sqrt{\bar{\zeta}_j}\right)}\right]\prod_{\zeta_j<0}\left(\zeta - \sqrt{\zeta_j}\right),$$

we find that

$$\bar{\psi}(\zeta) = \frac{1}{\psi(\zeta)}\prod_{\zeta_j<0}\left(\zeta^2 - \zeta_j\right).$$

Therefore, using the fact that $|\varphi(-1)| = 1$, we have

$$\bar{A}_0(z) = e^{g(z)/2}\bar{P}(z)\frac{\left(\psi(\sqrt{\zeta}) + \psi(-\sqrt{\zeta})\right)\varphi(-1)Q(z)(1-\zeta)^{\nu}}{\psi(\sqrt{\zeta})\psi(-\sqrt{\zeta})2\sqrt{\varphi(-1)}}.$$

With (21) taken into account, we get

$$\bar{A}_0(z) = e^{g(z)/2}\frac{\bar{P}(z)}{\chi(z)}\,\frac{\psi(\sqrt{\zeta}) + \psi(-\sqrt{\zeta})}{2\psi(\sqrt{\zeta})\psi(-\sqrt{\zeta})(1-\zeta)^{\nu}} = A_0(z),$$

because $\bar{P}(z)/\chi(z) = P(z)$. It can be verified similarly that $\bar{B}_0(z) = B_0(z)$.

c) We prove that $(z+1)B_0(z)/A_0(z) \in \mathscr{S}(E)$. By (22), to do this it suffices to prove that the class $\mathscr{S}$ contains the function

$$i\frac{\psi(\sqrt{\zeta}) + \psi(-\sqrt{\zeta})}{\sqrt{\zeta}\left(\psi(\sqrt{\zeta}) - \psi(-\sqrt{\zeta})\right)} \qquad \left(\zeta = \frac{z-1}{z+1}\right),$$

and this, in turn, is equivalent to the class $\mathscr{R}$ containing the function

$$i\frac{(\psi(\zeta)/\psi(-\zeta))+1}{(\psi(\zeta)/\psi(-\zeta))-1}.$$

The latter is obvious, since $|\psi(\zeta)/\psi(-\zeta)|<1$ when $\operatorname{Im}\zeta>0$. The theorem is proved.

§6. Description of all the $\mathscr{S}(E)$-pairs giving a representation of a function f

LEMMA. *Each pair of real entire functions X and Y satisfying Pell's functional equation*

$$X^2(z)+(z-a)(z-b)Y^2(z)=1$$

and the condition

$$(z-a)Y(z)/X(z)\in\mathscr{S}(E),$$

has the form

$$X(z)=\cos\left(\gamma\sqrt{(z-a)(z-b)}\right),$$
$$Y(z)=\sin\left(\gamma\sqrt{(z-a)(z-b)}\right)\Big/\sqrt{(z-a)(z-b)},$$

where $\gamma>0$.

PROOF. Without loss of generality it will be assumed that $a=-1$ and $b=1$. Let X, Y be one of the pairs under consideration. Setting $z=\cosh\zeta$, we arrive at the identity

$$X^2(\cosh\zeta)+\sinh^2\zeta\, Y^2(\cosh\zeta)=1,$$

which implies that the function G defined by

$$X(\cosh\zeta)+i\sinh\zeta\, Y(\cosh\zeta)=e^{iG(\zeta)}, \tag{23}$$

has the following properties:

1°. G is an entire function.

2°. $G(\zeta)$ is real for real ζ.

3°. $G(\zeta)$ is real for $\zeta=x+i\pi$ $(x=\bar{x})$.

4°. G is an odd function.

By (23),

$$X(\cosh\zeta)=\cos G(\zeta),\qquad Y(\cosh\zeta)=\sin G(\zeta)/\sinh\zeta$$

or

$$X(z)=\cos G\left(\ln\left(z+\sqrt{z^2-1}\right)\right),\qquad Y(z)=\sin G\left(\ln\left(z+\sqrt{z^2-1}\right)\right)\Big/\sqrt{z^2-1}. \tag{24}$$

Since $(z+1)Y(z)/X(z)\in\mathscr{S}(E)$, it follows (see §1) that

$$\frac{z-1}{z+1}\left(\frac{z^2+1}{2z}+1\right)Y\left(\frac{z^2+1}{2z}\right)\Big/X\left(\frac{z^2+1}{2z}\right)\in\mathscr{R},$$

i.e., $\tan H(z) \in \mathscr{R}$, where $H(z) = G(\ln z)$. The function H is holomorphic in the upper half-plane and, since

$$\operatorname{Im} \tan H(z) = \frac{\sinh[2 \operatorname{Im} H(z)]}{2|\cos H(z)|^2} \geqslant 0 \quad \text{for } \operatorname{Im} z > 0,$$

we have that $\operatorname{Im} H(z) \geqslant 0$ when $\operatorname{Im} z > 0$, i.e., $H \in \mathscr{R}$.

By property 2° of G, the function H is holomorphic and real for $z > 0$, and by property 3° it is holomorphic and real for $z < 0$. Therefore, $d\sigma(t) = 0$ for $t \neq 0$ in (5), and, consequently,

$$H(z) = \alpha + \beta z - \rho/z,$$

where $\alpha = \bar{\alpha}$, $\beta \geqslant 0$ and $\rho \geqslant 0$.

Property 4° gives us that $H(1/z) = -H(z)$, which implies that $\alpha = 0$ and $\beta = \rho$. Therefore,

$$G(\ln z) = \rho(z - 1/z) \quad \text{and} \quad G\left(\ln\left(z + \sqrt{z^2 - 1}\right)\right) = 2\rho\sqrt{z^2 - 1}.$$

Using the last equality in (24), we get the lemma.

An $\mathscr{S}(E)$*-pair* for f is defined to be a pair of real entire functions A, B that have no common zeros, realize the representation (19), and satisfy the condition $(z - a)B(z)/A(z) \in \mathscr{S}(E)$.

THEOREM 6. *The set of all* $\mathscr{S}(E)$*-pairs for an entire function* f *positive on* $E = (-\infty, \infty)\setminus(a, b)$ *is described by*

$$\begin{aligned} A(z) &= A_0(z)\cos\left(\gamma\sqrt{(z-a)(z-b)}\right) \\ &\quad - \sqrt{(z-a)(z-b)}\, B_0(z)\sin\left(\gamma\sqrt{(z-a)(z-b)}\right), \\ B(z) &= A_0(z)\sin\left(\gamma\sqrt{(z-a)(z-b)}\right)\Big/\sqrt{(z-a)(z-b)} \\ &\quad + B_0(z)\cos\left(\gamma\sqrt{(z-a)(z-b)}\right), \end{aligned} \tag{25}$$

where γ *runs through the set of nonnegative numbers.*

PROOF. We again assume that $a = -1$ and $b = 1$. Let us verify first of all that the formulas (25) give an $\mathscr{S}(E)$-pair for f for each $\gamma > 0$. The equality $A^2(z) + (z^2 - 1)B^2(z) = f(z)$ can be verified directly. We prove that

$$(z + 1)B(z)/A(z) \in \mathscr{S}(E).$$

This follows from the fact that the ratio can be represented in the form

$$\frac{(z+1)B(z)}{A(z)} = \frac{u(z) + v(z)}{1 - [(z-1)/(z+1)]\,u(z)v(z)} \tag{26}$$

where

$$u(z) = \sqrt{\frac{z-1}{z+1}}\, \tan\left(\gamma\sqrt{z^2-1}\right) \in \mathscr{S}(E), \quad v(z) = \frac{(z+1)B_0(z)}{A_0(z)} \in \mathscr{S}(E)$$

(see §1).

Conversely, suppose that A and B make up an $\mathscr{S}(E)$-pair for f. Let X and Y be determined from

$$\begin{aligned} A(z) &= A_0(z)X(z) - (z^2-1)B_0(z)Y(z), \\ B(z) &= B_0(z)X(z) + A_0(z)Y(z), \end{aligned} \tag{27}$$

so that

$$\begin{aligned} X(z) &= \big(A(z)A_0(z) + (z^2-1)B(z)B_0(z)\big)/f(z), \\ Y(z) &= \big(A_0(z)B(z) - A(z)B_0(z)\big)/f(z). \end{aligned} \tag{28}$$

The functions X and Y are real and satisfy Pell's functional equation

$$X^2(z) + (z^2-1)Y^2(z) = 1.$$

We prove that X and Y are entire functions. Since the functions $(z+1)B(z)/A(z)$, $(z+1)B_0(z)/A_0(z)$, and $A(z)/(1-z)B(z)$ serve as solutions of Problem $(\mathscr{S}(E); \sqrt{(z-a)/(b-z)}\,)$ associated with the roots of f, each root of f is a root of multiplicity at least as great as the difference between these solutions:

$$\frac{(z+1)B_0(z)}{A_0(z)} - \frac{A(z)}{(1-z)B(z)} = \frac{A(z)A_0(z) + (z^2-1)B(z)B_0(z)}{(1-z)A_0(z)B(z)},$$

$$\frac{(z+1)B(z)}{A(z)} - \frac{(z+1)B_0(z)}{A_0(z)} = \frac{(z+1)\big(A_0(z)B(z) - A(z)B_0(z)\big)}{A_0(z)A(z)}.$$

Thus, (28) defines entire functions X and Y.

To complete the proof of the theorem it remains to show that $(z+1)Y(z)/X(z) \in \mathscr{S}(E)$, after which the lemma can be used. As already noted, $(z+1)B(z)/A(z)$ is one of the solutions of Problem $(\mathscr{S}(E); \sqrt{(1+z)/(1-z)}\,)$ for the roots of f in the upper half-plane or $(-1, 1)$. If the formulas (22) are compared with (8), then it turns out that a description of all the solutions of this problem can be obtained by the formula

$$F(z) = \frac{A_0(z)\Omega(z) + (2/(1-\zeta))B_0(z)}{-(2\zeta/(1-\zeta))B_0(z)\Omega(z) + A_0(z)} \qquad \left(\zeta = \frac{z-1}{z+1}\right),$$

i.e., by the formula

$$F(z) = \frac{A_0(z)\Omega(z) + (z+1)B_0(z)}{-(z-1)B_0(z)\Omega(z) + A_0(z)},$$

where Ω runs through $\mathscr{S}(E)$. From (27) we see that $F(z) = (z+1)B(z)/A(z)$ is obtained for $\Omega(z) = (z+1)Y(z)/X(z)$, i.e., $(z+1)Y(z)/X(z) \in \mathscr{S}(E)$, which is what was required to prove.

REFERENCES

1. M. G. Kreĭn, *On the representation of functions by Fourier-Stieltjes integrals*, Uchen. Zap. Kuĭbyshev. Gos. Ped. i Uchit. Inst. **7** (1943), 123–147. (Russian)

2. N. I. Akhiezer, *On the theory of entire functions of finite degree*, Dokl. Akad. Nauk SSSR **63** (1948), 475–478. (Russian)

3. N. G. Chebotarev and N. N. Meĭman, *The Routh-Hurwitz problem for polynomials and entire functions*, Trudy Mat. Inst. Steklov. **26** (1949). (Russian)

4. N. I. Akhiezer, *Lectures on the theory of approximation*, 2nd ed., "Nauka", Moscow, 1965; English transl. of 1st ed., Ungar, New York, 1956.

5. B. Ya. Levin, *Distribution of zeros of entire functions*, GTTI, Moscow, 1956; English transl., Amer. Math. Soc., Providence, R.I., 1964.

6. Louis de Branges, *Hilbert spaces of entire functions*, Prentice-Hall, Englewood Cliffs, N.J., 1968.

7. M. G. Kreĭn and A. A. Nudel'man, *On direct and inverse problems for the boundary dissipation frequencies of a nonuniform string*, Dokl. Akad. Nauk SSSR **247** (1979), 1046–1049; English transl. in Soviet Math. Dokl. **20** (1979).

8. A. M. Rybalko, *A certain representation of entire functions of class A that are positive on the real axis*, Teor. Funktsiĭ, Funktsional. Anal. i Prilozhen. Vyp. 7 (1968), 53–58. (Russian)

9. N. H. Abel, *Über die Integration der Differential-Formel* $\rho\,dx/\sqrt{R}$, *wenn R und ρ ganze Functionen sind*, J. Reine Angew. Math. **1** (1826), 185–221; French transl. in his *Oeuvres complètes*. Vol. 1, Gröndahl, Christiania, 1839, pp. 33–65; rev. ed., 1881, pp. 104–144.

10. M. G. Kreĭn, *On inverse problems in the theory of filters and λ-zones of stability*, Dokl. Akad. Nauk SSSR **93** (1953), 767–770. (Russian)

11. V. A. Marchenko and I. V. Ostrovskiĭ, *A characterization of the spectrum of Hill's operator*, Mat. Sb. **97 (139)** (1975), 540–606; English transl. in Math. USSR Sb. **26** (1975).

12. N. I. Akhiezer, *On a generalization of the Fourier transform and the Paley-Wiener theorem*, Dokl. Akad. Nauk SSSR **96** (1954), 889–892. (Russian)

13. V. P. Gurariĭ, *The Fourier transformation of the space* $L^2(-\infty,\infty)$ *with a weight*, Mat. Sb. **58 (100)** (1962), 439–452. (Russian)

14. A. M. Rybalko, *On the Fourier transformation in the space* $L^2(-\infty,\infty)$ *with a weight*, Trudy Fiz.-Tekh. Inst. Nizkikh Temperatur Akad. Nauk Ukrain. SSR Mat. Fiz. i Funktsional. Anal. Vyp. 1 (1969), 147–164. (Russian)

15. N. I. Akhiezer, Remark on [14], ibid., pp. 165–171. (Russian)

16. M. G. Kreĭn and A. A. Nudel'man, *On some propositions of the type of the Fourier-Plancherel and Paley-Wiener theorems that can be obtained by methods in the theory of spectral functions*, Funktsional. Anal. i Prilozhen. **23** (1979), no. 4, 79–80; English transl. in Functional Anal. Appl. **23** (1979).

17. I. S. Kats and M. G. Kreĭn, *R-functions — analytic functions mapping the upper half-plane into itself*, Supplement I to the Russian transl. of F. V. Atkinson, *Discrete and continuous boundary problems*, "Mir", Moscow, 1968, pp. 629–647; English transl. in Amer. Math. Soc. Transl. (2) **103** (1973).

18. M. G. Kreĭn and A. A. Nudel'man, *The Markov moment problem and extremal problems*, "Nauka", Moscow, 1973; English transl., Amer. Math. Soc., Providence, R.I., 1977.

19. M. G. Kreĭn, *Concerning a special class of entire and meromorphic functions*, Paper #2 in N. I. Akhiezer and M. G. Kreĭn, *Some questions in the theory of moments*, Gos. Nauchno-Tekhn. Izdat. Ukraïn, Kharkov, 1938; English transl., Amer. Math. Soc., Providence, R.I., 1962.

Translated by H. H. MCFADEN

Amer. Math. Soc. Transl.
(2) Vol. **127**, 1986

On Singular Perturbations in Optimal Control Problems*

UDC 519.3

S. G. KREĬN AND G. A. KURINA

Let us consider a perturbed equation of state of the form

$$(A + \varepsilon B)\dot{x}(t) = Cx(t) + Du(t),\ x(0) = x^0, \tag{1}$$

where $x(t) \in R$, $m > 1$, $t \in [0, T]$, $\varepsilon \in (0, \varepsilon_1]$, $\varepsilon_1 > 0$, $u(t) \in R^m$, and A, B, C, and D are constant matrices, the matrix A being singular. We shall assume that the matrices $A + \varepsilon B$ and $A + \mu C$ are invertible for sufficiently small nonzero values of ε and μ respectively. Suppose that the kernel of the matrix A is one-dimensional. We shall denote by e_1 the null element of the matrix A, and by $e_2, \ldots, e_n$ a chain of all the elements adjoined to the former. It is assumed that the matrix A has the form $A = \left(\begin{smallmatrix} J & 0 \\ 0 & I \end{smallmatrix}\right)$, where J is a Jordan block with zero eigenvalue, and I is the unit matrix. The inner product $\langle \cdot\,, \cdot \rangle$ in R^m is defined in the usual way, as a sum of products of the corresponding variables.

The perturbed equation (1) is a system of differential equations in the variables $x^1(t), \ldots, x^m(t)$. The solution of equation (1) for $\varepsilon = 0$ will be marked with a bar. We shall assume that the algebraic equation obtained from (1) for $\varepsilon = 0$ can be uniquely solved with respect to $\bar{x}^1(t)$.

At first we shall study the nonclassical Mayer problem. With the aid of controls $u(t)$ that are bounded and measurable r-dimensional vector-valued functions such that

$$|u^i(t)| \leqslant 1, \qquad t \in [0, T],\ i = 1, \ldots, r, \tag{2}$$

it is required to transfer the system (1) from its initial position during a specified time T to any final position in such a way that the following functional is minimized:

$$I(u) = \langle d, x(T) \rangle. \tag{3}$$

Here d is a fixed vector in R^m.

1980 *Mathematics Subject Classification*. Primary 49B50; Secondary 49E25.

*Translation of Stability of Motion. Analytical Mechanics. Control of Motion (V. M. Matrosov and V. G. Demin, editors), "Nauka", Moscow, 1981, pp. 170–178; MR **83j**: 49020.

We shall study the behavior of an optimal control and of an optimal path of problem (1)–(3) as $\varepsilon \to 0$.

For the case that the matrix A does not have elements that are adjoined to the eigenelement that corresponds to a zero eigenvalue, i.e., for $A = \left(\begin{smallmatrix} 0 & 0 \\ 0 & I \end{smallmatrix}\right)$ and a matrix B of the form $\left(\begin{smallmatrix} I & 0 \\ 0 & 0 \end{smallmatrix}\right)$, we can find problems similar to (1)–(3) in references [1]–[4].

Let us write down the equation for the conjugate variable $\varphi(t)$ in problem (1)–(3), i.e.,

$$\dot{\varphi}(t) = -\left((A + \varepsilon B)^{-1} C\right)' \varphi(t) = -C'(A' + \varepsilon B')^{-1} \varphi(t), \tag{4}$$

$$\varphi(T) = d. \tag{5}$$

Let us introduce a new function

$$\psi(t, \varepsilon) = (A' + \varepsilon B')^{-1} \varphi(t). \tag{6}$$

Then (4) and (5) can be written as follows:

$$(A' + \varepsilon B')\dot{\psi}(t, \varepsilon) = -C'\psi(t, \varepsilon), \tag{7}$$

$$\psi(T, \varepsilon) = -(A' + \varepsilon B')^{-1} d. \tag{8}$$

In [5] it is shown that for a matrix $A + \varepsilon B$ to be invertible, it is necessary and sufficient that there exist a number $p \geqslant 1$ such that

$$\left\langle B(A'B)^i e_1, e_n \right\rangle = 0, \quad i = 0, \dots, p-2, \qquad \left\langle B(A'B)^{p-1} e_1, e_n \right\rangle \neq 0.$$

Then the Laurent expansion of the matrix $(A' + \varepsilon B')^{-1}$ in terms of ε in a neighborhood of the origin will be

$$(A' + \varepsilon B')^{-1} = \frac{R_{-p}}{\varepsilon^p} + \frac{R_{-p+1}}{\varepsilon^{p-1}} + \cdots + \frac{R_{-1}}{\varepsilon} + R_0 + \varepsilon R_1 + \cdots; \qquad R_{-p} \neq 0.$$

Using the technique described in [13], [5] and [6] let us construct an asymptotic expansion of the solution $\psi(t, \varepsilon)$ of problem (7), (8) in powers of ε. This expansion will be sought in the form

$$\psi(t, \varepsilon) = \sum_{k=-p}^{q} \varepsilon^k (\psi_k(t) + Q_k \psi(\tau_1)) + r_q \psi(t, \varepsilon), \tag{9}$$

where $\tau_1 = (t - T)/\varepsilon^p$, the $Q_k\psi(\tau_1)$ are boundary layer functions in a neighborhood of $t = T$, and $r_q\psi(t, \varepsilon)$ has bounds

$$\| r_q \psi(t, \varepsilon) \| \leqslant c\varepsilon^{q+1}, \qquad \| \dot{r}_q \psi(t, \varepsilon) \| \leqslant c\varepsilon^{q+1-p}, \tag{10}$$

with a constant c that does not depend on ε or t.

For a function $\psi(t, \varepsilon)$ that has an expansion (9) with arbitrary $q \geqslant 0$ to be a solution of (7), it is necessary and sufficient that the functions $\psi_k(t)$ be solutions of the equations

$$A'\dot{\psi}_{-1}(t) = -C'\psi_{-p}(t),$$

$$A'\dot{\psi}_k(t) = -C'\psi_k(t) - B'\dot{\psi}_{k-1}(t), \qquad k = -p + 1, \dots, q, \tag{11}$$

the $Q_k\psi(\tau_1)$ be solutions of the equations

$$A'\frac{dQ_{-p}\psi(\tau_1)}{d\tau_1} = 0,$$

$$A'\frac{dQ_k\psi(\tau_1)}{d\tau_1} = -B'\frac{dQ_{k-1}\psi(\tau_1)}{d\tau_1}, \quad k = -p+1,\dots,-1 \quad \text{for } p > 1, \tag{12}$$

$$A'\frac{dQ_k\psi(\tau_1)}{d\tau_1} = -B'\frac{dQ_{k-1}\psi(\tau_1)}{d\tau_1} - C'Q_{k-p}\psi(\tau_1), \qquad k = 0,\dots,q,$$

and the remainder $r_q\psi(t, \varepsilon)$ be a solution of the equation

$$(A' + \varepsilon B')\dot{r}_q\psi(t,\varepsilon) = -C'r_q\psi(t,\varepsilon) - \varepsilon^{q+1}B'\dot{\psi}_q(t)$$
$$-\varepsilon^{q+1-p}B'\frac{dQ_k\psi(\tau_1)}{d\tau_1} - C'\sum_{k=-p+q+1}^{q}\varepsilon^k Q_k\psi(\tau_1). \tag{13}$$

The terminal conditions for the functions occurring in (9) will be assigned as follows:

$$\psi_k(T) + Q_k\psi(0) = -R_k d, \qquad k = p,\dots,q, \tag{14}$$

$$r_q\psi(T,\varepsilon) = -\varepsilon^{q+1}R_{q+1}d - \varepsilon^{q+2}R_{q+2}d - \cdots. \tag{15}$$

As in [5], we can find from (11), (12), and (14) all the functions $\psi_k(t)$, $k = -p,\dots,q$, as well as the functions $Q_k\psi(\tau_1)$, $k = -p,\dots,q-p$, with $\psi_k(t) \equiv 0$ for $k = -p,\dots,-1$. For finding the other functions $Q_k\psi(\tau_1)$, $k = q-p+1,\dots,q$, we must write $\psi(t,\varepsilon)$ in the form

$$\psi(t,\varepsilon) = \sum_{k=-p}^{q+p}\varepsilon^k(\psi_k(t) + Q_k\psi(\tau_1)) + r_{q+p}\psi(t,\varepsilon),$$

then consider (11), (12), and (14) for $k = -p,\dots,q+p$, and denote by $r_q\psi(t,\varepsilon)$ the expression

$$\sum_{k=q+1}^{q+p}\varepsilon^k(\psi_k(t) + Q_k\psi(\tau_1)) + r_{q+p}\psi(t,\varepsilon).$$

All the functions $Q_k\psi(\tau_1)$ will be boundary layer functions if and only if

$$\omega = \frac{\langle Ce_1, e_n\rangle}{(-1)^{p-1}\langle B(A'B)^{p-1}e_1, e_n\rangle} < 0.$$

Here we have

$$\|Q_k\psi(\tau_1)\| \leqslant P_k(\tau_1)\exp(-\omega\tau_1),$$
$$\|dQ_k\psi(\tau_1)/d\tau_1\| \leqslant \tilde{P}_k(\tau_1)\exp(-\omega\tau_1),$$

$k = -p,\dots,q$, $P_k(\tau_1)$ and $\tilde{P}_k(\tau_1)$ being polynomials in τ_1.

The function $\psi_0(t)$ can be determined from the singular equation (7),

$$A'\dot{\psi}_0(t) = -C'\psi_0(t) \tag{16}$$

with a condition

$$\psi_0(T) = -Ad + \frac{\langle C'Ad, e_1\rangle}{\langle C'e_n, e_1\rangle} e_n + \frac{\langle d, e_1\rangle}{\langle C'e_n, e_1\rangle}\left[AC'e_n - \frac{\langle C'AC'e_n, e_1\rangle}{\langle C'e_n, e_1\rangle} e_n\right]. \tag{17}$$

It is possible to prove that a remainder $r_q\psi(t, \varepsilon)$ which is a solution of problem (13), (15) has bounds (10). Thus we have

LEMMA 1. *Under the condition* $\omega < 0$ *the solution* $\psi(t, \varepsilon)$ *of problem* (7), (8) *will have an asymptotic expansion* (9). *As* $\varepsilon \to 0$ *the function* $\psi(t, \varepsilon)$ *will tend to the solution of problem* (16), (17) *uniformly in* $t \in [0, T_1]$, $0 < T_1 < T$.

For $\varepsilon = 0$ we obtain from (1) a singular problem of dimension $m - 1$ for the variables x^i, $i = 2, \ldots, m$. Let us assume that it makes sense and that the dimension of the control vector does not change in the case of singularity. The system for the conjugate variables $\bar\psi^1, \ldots, \bar\psi^{n-1}, \bar\psi^{n+1}, \ldots, \bar\psi^m$ for the case of a singular problem can be written in the form $A'\dot{\bar\psi}(t) = C'\bar\psi(t)$ if we assume that

$$\bar\psi^n = -\frac{1}{c_{n1}} \sum_{\substack{j=1 \\ j\neq n}}^{m} c_{j1}\bar\psi^j, \qquad C = (c_{ij}).$$

Let us write Pontryagin's function $H = \langle \varphi, (A + \varepsilon B)^{-1}(Cx + Du)\rangle$ for a perturbed problem [7]. By virtue of (6) we have

$$H = \langle \psi, Cx + Du\rangle.$$

Pontryagin's function H for a singular problem has the same form as for a perturbed problem, i.e., $H = \langle \bar\psi, C\bar x + Du\rangle$.

Let us assume that for almost all $t \in [0, T]$ the maximum with respect to u in (2) of the function $\langle \psi(t, \varepsilon), Du\rangle$, where $\psi(t, \varepsilon)$ is a nontrivial solution of (7), is attained for a unique u in (2) for sufficiently small $\varepsilon \geqslant 0$. For $\varepsilon = 0$ we shall assume that $\psi(t, 0) = \psi_0(t)$. It is also assumed that $\psi_0(T) \neq 0$.

Since the optimal controls for a perturbed problem and a singular problem can be determined from Pontryagin's maximum principle [7], [12], we can prove with the aid of Lemma 1 and of the results of [8] the following assertion:

THEOREM 1. *As* $\varepsilon \to 0$ *the optimal control* $u(t, \varepsilon)$ *of problem* (1)–(3) *tends, in the norm of* $L_1[0, T]$, *to the optimal control* $\bar u(t)$ *of a singular problem with the same controls from* (2) *and with a performance criterion*

$$\bar J(\bar u) = -\sum_{j=1}^{n-1} \bar x^{j+1}(T)\bar\psi^j(T) - \sum_{j=n+1}^{m} \bar x^j(T)\bar\psi^j(T),$$

where $\bar\psi(T) = \psi_0(T)$ *and* $\bar x^j(0) = x^j(0)$, $j = 2, \ldots, m$.

By virtue of condition (2) it follows that convergence in $L_1[0, T]$ implies convergence in any $L_q[0, T]$, $1 \leqslant q < \infty$.

The equation for the optimal path $x(t, \varepsilon)$ of the perturbation problem will be written as follows:

$$(A_\mu + \varepsilon B_\mu)\dot{x}(t, \varepsilon) = \frac{1}{\mu}x(t, \varepsilon) - \frac{1}{\mu}A_\mu x(t, \varepsilon) + D_\mu u(t, \varepsilon), \tag{18}$$

where $A_\mu = (A + \mu C)^{-1}A$, $B_\mu = (A + \mu C)^{-1}B$ and $D_\mu = (A + \mu C)^{-1}D$.

According to [5] the space R^m can be decomposed into a direct sum of the root subspace N_1 of the matrix A_μ and of the subspace $M_1 = \{x \in R^m: \langle Cx, e_n\rangle = 0\}$ which are invariant with respect to the operator A_μ, with A_μ having an inverse on M_1. By P_1 we denote the projection operator of R^m onto N_1, and by Q_1 the projection operator of R^m onto M_1.

The function $x(t, \varepsilon)$ will be written in the form $x(t, \varepsilon) = Q_1 x(t, \varepsilon) + y(t, \varepsilon)e_1$, where $y(t, \varepsilon)e_1 = P_1 x(t, \varepsilon)$, with $y(t, \varepsilon)$ being a scalar function. Equation (18) can be transformed into a system of differential equations for the functions $Q_1 x(t, \varepsilon)$ and $y(t, \varepsilon)$.

Let us introduce a function $\tilde{x}(t, \varepsilon)$ which is a solution of (1) for $u(t) = \tilde{u}(t)$. By comparing $x(t, \varepsilon)$ with $\tilde{x}(t, \varepsilon)$, and $\tilde{x}(t, \varepsilon)$ with the solution $\bar{x}(t)$ of (1) for $\varepsilon = 0$, $u(t) = \tilde{u}(t)$, and $Q_1 x(0) = Q_1 x^0$, we obtain

LEMMA 2. *As* $\varepsilon \to 0$

$$\| y(t, \varepsilon) - \bar{y}(t) \|_{L_1[0,T]} \to 0, \qquad \|Q_1 x(t, \varepsilon) - Q_1 \bar{x}(t)\|_{L_1[0,T]} \to 0,$$

where $\bar{y}(t)e_1 = P_1\bar{x}(t)$.

Let us note that $|y(t, \varepsilon)|$ is uniformly bounded with respect to $t \in [0, T]$ and with respect to ε in a neighborhood of the origin.

From the previous lemma we obtain

THEOREM 2. *If* $x(t, \varepsilon) = (x^1(t, \varepsilon), \dots, x^m(t, \varepsilon))$ *is an optimal path of the perturbation problem* (1)–(3), *and* $(x^2(t), \dots, x^m(t))$ *is an optimal path of a singular problem with a criterion* $\bar{J}(\bar{u})$, *where*

$$\bar{x}^1(t) = -\left[\sum_{j=2}^{m} c_{nj}\bar{x}^j(t) + \langle D\bar{u}(t), e_n\rangle\right]\Big/ c_{n1},$$

$$\bar{x}^j(0) = x^j(0, \varepsilon), \qquad j = 2, \dots, m,$$

then

$$\|x^j(t\varepsilon) - \bar{x}^j(t)\|_{L_1[0,T]} \to 0, \qquad j = 2, \dots, m$$

and

$$\|x^1(t, \varepsilon) - \bar{x}^{-1}(t)\|_{L_1[0,T]} \to 0 \quad \textit{as } \varepsilon \to 0.$$

Next we consider a classical problem without constraints on the controls. With the aid of continuous controls $u(t)$ it is required to transfer the system (1) from its initial position during an assigned time T to an arbitrary final position, with minimization of the functional

$$J(u) = \langle d, x(T)\rangle + \frac{1}{2}\int_0^T \langle u(t), Ru(t)\rangle\, dt. \tag{19}$$

Here d is a fixed vector in R^m, with R being an $r \times r$ positive definite matrix. We study the behavior of the solution of problem (1), (19) as $\varepsilon \to 0$.

For the case $A = \left(\begin{smallmatrix} 0 & 0 \\ 0 & I \end{smallmatrix}\right)$ and $B = \left(\begin{smallmatrix} I & 0 \\ 0 & 0 \end{smallmatrix}\right)$ such a problem has been studied in [9] under more general assumptions concerning the state equations and the performance criterion, but with $d = 0$. The results of that paper have been extended by Dmitriev [10]. This problem has been considered in [1] with exactly the same performance criterion (19), but with state equations $\dot{x}(t) = A_1x(t) + B_1y(t) + C_1u(t)$ and $\varepsilon\dot{y}(t) = A_2x(t) + B_2y(t)$, where the matrices A_1, B_1, C_1, A_2, and B_2 are constant. Classical singularly perturbed problems of optimal control have been studied also in [4] and [11].

The conjugate variable $\varphi(t)$ of the problem (1), (19) satisfies equation (4) with a terminal condition (5). From Pontryagin's maximum principle [7], [12] it follows that

$$\partial K/\partial u = 0, \tag{20}$$

where

$$K = \left\langle \varphi, (A + \varepsilon B)^{-1}Cx + (A + \varepsilon B)^{-1}Du\right\rangle - 1/2\langle u, Ru\rangle$$

is the Hamiltonian of problem (1), (19). From (20) we can find an optimal control; it is unique and has the form $u(t, \varepsilon) = R^{-1}D'\psi(t, \varepsilon)$. Here $\psi(t, \varepsilon)$ is specified by (6).

Since we have constructed an asymptotic expansion of $\psi(t, \varepsilon)$ in powers of ε, it follows that there exists also an expansion of an optimal control of problem (1), (19) in powers of ε.

Let us substitute $u(t, \varepsilon)$ into (1). We denote $DR^{-1}D'$ by S. For an optimal path of problem (1), (19) we obtain the equation

$$(A + \varepsilon B)\dot{x}(t, \varepsilon) = Cx(t, \varepsilon) + S\psi(t, \varepsilon), \qquad x(0, \varepsilon) = x^0. \tag{21}$$

With the aid of the methods developed in [5], [6] and [13] let us construct an asymptotic expansion of $x(t, \varepsilon)$ in powers of ε. This expansion will be sought in the form

$$x(t, \varepsilon) = \sum_{k=-p}^{q} \varepsilon^k(x_k(t) + \Pi_k x(\tau_0) + Q_k x(\tau_1)) + r_q x(t, \varepsilon), \tag{22}$$

where $\tau_0 = t/\varepsilon^p$, the $\Pi_k x(\tau_0)$ are boundary layer functions in a neighborhood of $t = 0$, the $Q_k x(\tau_1)$ are boundary layer functions in a neighborhood of $t = T$, and $r_q x(t, \varepsilon)$ for $t \in [0, T]$ has bounds

$$\left\| r_q x(t, \varepsilon) \right\| \leqslant c\varepsilon^{q+1}, \qquad \left\| \dot{r}_q x(t, \varepsilon) \right\| \leqslant c\varepsilon^{q+1-p} \tag{23}$$

with a constant c that does not depend on ε or t.

Let $s > p$ and $\omega < 0$. For a function $x(t, \varepsilon)$ that has an expansion (22) with arbitrary $q \geqslant 0$ to be a solution of (21), it is necessary and sufficient that the functions $x_k(t)$ must be solutions of the equations

$$A\dot{x}_{-s}(t) = Cx_{-s}(t), \quad A\dot{x}_k(t) = Cx_k(t) - B\dot{x}_{k-1}(t), \quad k = -s+1, \ldots, -1,$$
$$A\dot{x}_k(t) = Cx_k(t) - B\dot{x}_{k-1}(t) + S\psi_k(t), \qquad k = 0, \ldots, q;$$

the functions $\Pi_k x(\tau_0)$ must be solutions of the equations

$$A\frac{d\Pi_{-s}x(\tau_0)}{d\tau_0} = 0, \qquad A\frac{d\Pi_k x(\tau_0)}{d\tau_0} = -B\frac{d\Pi_{k-1}x(\tau_0)}{d\tau_0},$$
$$k = -s+1, \dots, p-s-1 \quad \text{for } p > 1,$$

$$A\frac{d\Pi_k x(\tau_0)}{d\tau_0} = -B\frac{d\Pi_{k-1}x(\tau_0)}{d\tau_0} + C\Pi_{k-p}x(\tau_0), \qquad k = p-s, \dots, q;$$

the functions $Q_k x(\tau_1)$ must be solutions of the equations

$$A\frac{dQ_{-s}x(\tau_1)}{d\tau_1} = 0,$$

$$A\frac{dQ_k x(\tau_1)}{d\tau_1} = -B\frac{dQ_{k-1}x(\tau_1)}{d\tau_1}, \qquad k = -s+1, \dots, -s+p-1 \quad \text{for } p > 1,$$

$$A\frac{dQ_k x(\tau_1)}{d\tau_1} = -B\frac{dQ_{k-1}x(\tau_1)}{d\tau_1} + CQ_{k-p}x(\tau_1), \qquad k = -s+p, \dots, -1, \tag{24}$$

$$A\frac{dQ_k x(\tau_1)}{d\tau_1} = -B\frac{dQ_{k-1}x(\tau_1)}{d\tau_1} + CQ_{k-p}x(\tau_1) + SQ_{k-p}\psi(\tau_1), \quad k = 0, \dots, q;$$

and the remainder must satisfy the equation

$$\begin{aligned}(A+\varepsilon B)\dot{r}_q x(t,\varepsilon) &= Cr_q x(t,\varepsilon) + Sr_q\psi(t,\varepsilon) \\ &\quad - \varepsilon^{q+1}B\dot{x}_q(t) - \varepsilon^{q+1-p}B\left(\frac{d\Pi_q x(\tau_0)}{d\tau_0} + \frac{dQ_q x(\tau_1)}{d\tau_1}\right) \\ &\quad + \sum_{k=q+1-p}^{q} \varepsilon^k\left[C(\Pi_k x(\tau_0) + Q_k x(\tau_1)) + SQ_k\psi(\tau_1)\right].\end{aligned} \tag{25}$$

The initial conditions for the functions occurring in (22) will be assigned as follows:

$$x_k(0) + \Pi_k x(0) = 0 \quad \text{for } k = -s, \dots, q, \quad k \neq 0,$$

$$x_0(0) + \Pi_0 x(0) = x^0, \qquad r_q x(0,\varepsilon) = -\sum_{k=-s}^{q} \varepsilon^k Q_k x(-T/\varepsilon^p). \tag{26}$$

From the previous formulas we can obtain all the functions $x_k(t)$ and $\Pi_k x(\tau_0)$. Then $x_k(t) \equiv 0$ and $\Pi_k x(\tau_0) \equiv 0$ for $k = -s, \dots, -1$, and

$$\|\Pi_k x(\tau_0)\| \leqslant S_k(\tau_0)\exp(\omega\tau_0), \qquad \left\|\frac{d\Pi_k x(\tau_0)}{d\tau_0}\right\| \leqslant \tilde{S}_k(\tau_0)\exp(\omega\tau_0),$$

where $S_k(\tau_0)$ and $\tilde{S}_k(\tau_0)$ are polynomials in τ_0, and $k = 0, \dots, q$.

As in [13], the initial conditions for the functions $Q_k x(\tau_1)$ will be determined by the requirement that the $Q_k x(\tau_1)$ must be boundary layer functions in a neighborhood of $t = T$. From (24) we obtain the functions $Q_k x(\tau_1)$, with $Q_k x(\tau_1) \equiv 0$ for $k = -s, \ldots, -p-1$ and

$$\|Q_k x(\tau_1)\| \leqslant P_k(\tau_1)\exp(-\omega\tau_1), \qquad \left\|\frac{dQ_k x(\tau_1)}{d\tau_1}\right\| \leqslant \tilde{P}_k(\tau_1)\exp(-\omega\tau_1),$$

where $P_k(\tau_1)$ and $\tilde{P}_k(\tau_1)$ are polynomials in τ_1, and $k = -p, \ldots, q$.

Now we can prove that the remainder $r_q x(t, \varepsilon)$, which is a solution of problem (25), (26), has bounds (23) for $\omega < 0$.

Thus, we have obtained for the optimal path an asymptotic expression

$$x(t,\varepsilon) = \sum_{k=-p}^{-1} \varepsilon^k Q_k x(\tau_1) + \sum_{k=0}^{q} \varepsilon^k (x_k(t) + \Pi_k x(\tau_0) + Q_k x(\tau_1)) + r_q x(t,\varepsilon).$$

By setting $q = 0$ in this expansion, we find that $x(t, \varepsilon)$ tends to $x_0(t)$ as $\varepsilon \to 0$ uniformly in $t \in [T_2, T_3]$, where $0 < T_2 < T_3 < T$, and $x_0(t)$ is a solution of the equation $Ax_0(t) = Cx_0(t) + S\psi_0(t)$. Let us note that $x(t, \varepsilon)$ tends to $x_0(t)$ as $\varepsilon \to 0$ also in the metric of $L_q[0, T_3]$ for any q, $1 \leqslant q < \infty$.

For $\varepsilon = 0$ we obtain from (1) a singular problem of dimension $m - 1$ for the variables $\bar{x}^i(t)$, where $\bar{x}^i(0) = x^i(0)$, $i = 2, \ldots, m$. Let us consider the problem of minimizing the functional

$$\bar{J}(u) = -\sum_{j=1}^{n-1} x^{j+1}(T)\psi_0^j(T) - \sum_{j=n+1}^{m} \bar{x}^j(T)\psi_0^j(T) + \frac{1}{2}\int_0^T \langle u(t), Ru(t)\rangle\, dt \tag{27}$$

on the paths of a singular system. From Pontryagin's maximum principle we obtain an optimal control $\bar{u}(t)$ for the singular problem, i.e., $\bar{u}(t) = R^{-1}D'\psi_0(t)$. From the foregoing analysis we obtain

THEOREM 3. *If all the above conditions and assumptions hold, then there exist asymptotic expansions for the optimal control, the optimal path, and the optimum value of the functional of the optimal control problem* (1), (19), *and as* $\varepsilon \to 0$ *the optimal control of the perturbation problem tends uniformly in* $t \in [0, T_1]$, $0 < T_1 < T$, *to the optimal control of a singular problem with a performance criterion* (27), *whereas the optimal path of the perturbation problem tends to the optimal path of the singular problem uniformly in* $t \in [T_2, T_3]$, $0 < T_2 < T_3 < T$.

Let us note that in both these problems it is necessary, in order to obtain the functional of the limit problem, to replace $x^1(T)$ in the performance criterion by $\bar{x}^1(t)$, and then drop the term outside the integral that contains the control.

Bibliography

1. M. G. Dmitriev, Candidate's Dissertation, Dnepropetrovsk State Univ., Dnepropetrovsk, 1972. (Russian)

2. ______, *On the continuity of the solution of Mayer's problem with respect to singular perturbations*, Zh. Vychisl. Mat. i Mat. Fiz. **12** (1972), 788–791; English transl. in USSR Comput. Math. and Math. Phys. **12** (1972).

3. ______, *A property of singularly perturbed systems*, Mathematics and Mechanics, collection of scientific papers by doctoral students, Dnepropetrovsk. Gos. Univ., Dnepropetrovsk, 1972, pp. 202–206. (Russian)

4. V. A. Anikeeva (Esipova), Candidate's Dissertation, Moscow State Univ., Moscow, 1974. (Russian)

5. S. P. Zubova, Candidate's Dissertation, Voronezh, State Univ., Voronezh, 1973. (Russian)

6. ______, *Asymptotic solution of certain classes of singularly perturbed equations*, All-Union Conf. Asymptotic Methods in the Theory of Singularly Perturbed Differential and Integrodifferential Equations and Their Applications, Abstracts of Reports, "Ilim", Frunze, 1975, pp. 36–39. (Russian) R. Zh. Mat. **1975** #12Б885.

7. L. S. Pontryagin et al., *Mathematical theory of optimal processes*, 2nd ed., "Nauka", Moscow, 1969; English transls. of 1st ed., Wiley, 1962, and MacMillan, 1964.

8. B. N. Pshenichnyĭ, *On a pursuit problem*, Kibernetika (Kiev) **1967** no. 6, 54–66; English transl. in Cybernetics **3** (1967).

9. P. Sannuti and P. Kokotović, *Singular perturbation method for near optimum design of high-order non-linear systems*, Automatica–J. IFAC **5** (1969), 773–779.

10. M. G. Dmitriev, *On an asymptotic solution of an optimal control problem*, Differential Equations and Their Applications, No. 1, Dnepropetrovsk. Gos. Univ., Dnepropetrovsk, 1971, pp. 21–29. (Russian)

11. M. G. Dmitriev, V. A. Esipova and V. I. Chuev, *Passage to the limit in a singularly perturbed problem of optimal control*, Differential Equations and Their Applications, No. 2, Dnepropetrovsk. Gos. Univ., Dnepropetrovsk, 1973, pp. 40–45. (Russian)

12. E. B. Lee and L. Markus, *Foundations of optimal control theory*, Wiley, 1967.

13. A. B. Vasil′eva and V. F. Butuzov, *Asymptotic expansions of solutions of singularly perturbed equations*, "Nauka", Moscow, 1973. (Russian)

Translated by L. EBNER

Amer. Math. Soc. Transl.
(2) Vol. **127**, 1986

Sets of Uniqueness for Complete Orthonormal Systems and Fourier Integrals*

UDC 517.51

G. G. GEVORKYAN

Abstract. This paper investigates the question of the existence of sets of uniqueness of arbitrarily small positive measure, both for arbitrary complete orthonormal systems and for certain integral transforms.

Following [1] and [2], we introduce two definitions.

DEFINITION 1. Let $\{\varphi_n(x)\}_1^\infty$ be a complete orthonormal system in $L_2[0,1]$. A set $E \subset [0,1]$ is said to be a *set of uniqueness* for $\{\varphi_n(x)\}_1^\infty$ if the conditions $f(x) = 0$ for $x \in E$ and $\sum_1^\infty |\int f\varphi_n|^p < +\infty$ for some p, $0 < p < 2$, imply that $f(x)$ vanishes almost everywhere on $[0,1]$.

Similarly we can define sets of uniqueness for integral transforms.

DEFINITION 2. Let the kernel $K(x,t)$, $0 \leqslant x,\ t < +\infty$, have the property that for every function([1])

$$\underset{\sigma\to\infty}{\text{l.i.m.}} \int_0^\sigma K(x,t)f(t)\,dt = \hat{f}(x), \quad \text{where } f(x) \in L_2(0,+\infty), \tag{1}$$

$$\|f\|_2 = \|\hat{f}\|_2; \tag{2}$$

then a set $E \subset (0,+\infty)$ is said to be a *set of uniqueness* for $K(x,t)$ if the conditions $f(x) = 0$ for $x \in E$ and $\hat{f}(x) \in L_p(0,+\infty)$ for some p, $1 \leqslant p < 2$, imply that $f(x) = 0$ almost everywhere on $[0,+\infty)$.

We have the following theorems.

THEOREM 1. *For every complete orthonormal system $\{\varphi_n(x)\}_1^\infty$ in $L_2[0,1]$ and for every $\varepsilon > 0$ there is a measurable set $E \subset [0,1]$, $\mu E < \varepsilon$, which is a set of uniqueness for $\{\varphi_n(x)\}_1^\infty$.*

1980 *Mathematics Subject Classification*. Primary 42C25, 42A38, 44A05; Secondary 42A63.

*Translation of Akad. Nauk Armyan. SSR Dokl. **72** (1981), 218–223; MR**82k:** 42022.

([1])Here $\text{l.i.m.}_{\sigma\to\infty}$ denotes the limit in the sense of convergence in $L_2(0,+\infty)$ as $\sigma \to \infty$, and $\|f\|_p = \{\int_0^\infty |f(x)|^p\,dx\}^{1/p}$.

THEOREM 2. *Let the kernel $K(x,t)$ satisfy conditions* (1) *and* (2), *and in addition let it be bounded in every strip* $(0,+\infty)\times(0,a)$, *i.e. for every $a>0$ let there be an $M(a)$ such that*

$$|K(x,t)|<M(a)<\infty \quad \text{when } 0\leqslant t\leqslant a,\, 0\leqslant x<+\infty. \tag{3}$$

Then for every $\varepsilon>0$ there is a measurable set $E\subset(0,+\infty)$, $\mu E<\varepsilon$, which is a set of uniqueness for $K(x,t)$.

Theorem 1 generalizes a theorem of Colzani [1], which establishes a similar result in the special case when the functions $\{\varphi_n(x)\}$ are uniformly bounded.

The next theorem follows immediately from Theorem 2.

THEOREM 3. *For every $\varepsilon>0$ there is a measurable set $E\subset(-\infty,+\infty)$, $\mu E<\varepsilon$, which has the following properties*:

1) *If $f\in L_2(-\infty,+\infty)$, $f(x)=0$ for $x\in E$, and $\hat f(x)=\frac{1}{2\pi}\int_{-\infty}^{+\infty}f(t)e^{-itx}\,dt\in L_p(-\infty,+\infty)$ for some p, $1\leqslant p<2$, then $f(x)=0$ almost everywhere on $(-\infty,+\infty)$.*

2) *If $f(x)\in L_p(-\infty,+\infty)\cap L_2(-\infty,+\infty)$ for some p, $1\leqslant p<2$, and $\hat f(x)=0$ for $x\in E$, then $f(x)=0$ almost everywhere on $(-\infty,+\infty)$.*

Theorems 1 and 2 are proved by using the following lemmas.

LEMMA 1. *Let $\{\varphi_n(x)\}_1^\infty$ be an arbitrary complete orthonormal system in $L_2(0,1)$. Then for every $\varepsilon>0$, $q>2$, and $(0,a)\subset(0,1)$ there are a measurable set $E_\varepsilon\subset(0,a)$ and a bounded function $f(x)$ such that*

1) *$f(x)=0$ when $x\in(a,1)$,*

2) *$f(x)=1$ when $x\in(0,a)\setminus E_\varepsilon$, $\mu E_\varepsilon<\varepsilon$, and*

3) $\{\sum_{n=1}^\infty|\int_0^1 f(x)\varphi_n(x)\,dx|^q\}^{1/q}<\varepsilon$.

LEMMA 2. *Let $K(x,t)$ be a kernel satisfying conditions* (1), (2), *and* (3). *Then for every $a\in(0,+\infty)$, $a>\varepsilon>0$, and $q>2$ there are a function $f(x)\in L_2(0,+\infty)$ and a set $E_\varepsilon\subset(0,a)$ such that*

1) *$f(x)=0$, $x\in(a,+\infty)$,*

2) *$f(x)=1$, $x\in(0,a)\setminus E_\varepsilon$, $\mu E_\varepsilon<\varepsilon$; and*

3) $\|\hat f(x)\|_q<\varepsilon$.

These lemmas can be proved by using lemmas established by Talalyan [3], [4] (see [3], Lemma 2, also [4], Lemma 1, and [5], Lemmas 2.11.3 and 2.11.4).

Here we present the proof of Theorem 2. The proofs of Theorem 1 and Lemma 1 are similar to the proofs of Theorem 2 and Lemma 2, given below.

Lemma 2 is proved by using the following lemma, whose proof is in effect given in [4] (see Lemma 1).

LEMMA 3. *Let $K(x,t)$ be a kernel satisfying conditions* (1) *and* (2), *$[a,b]\subset(0,+\infty)$, and let $A\subset(0,+\infty)$ be a measurable set of finite measure. Then for arbitrary ε, $1>\varepsilon>0$, $\alpha>0$, and $\beta>0$ there is a set $E\subset[a,b]$, $\mu E=\varepsilon(b-a)$,* *such that* $f(x) = \chi_{[a,b]}(x) - \chi_E(x)/\varepsilon$ *satisfies the condition*

$$\mu\{x \in A: |\hat{f}(x)| > \alpha\} < \beta.$$

PROOF OF LEMMA 2. We partition the interval $(0, a)$ into n equal parts $I_1, \dots, I_n$ (n will be chosen later). Lemma 3 implies the existence of a function $f(x)$ such that

$$f_1(x) = 0, \qquad x \notin I_1; \tag{4}$$

$$f_1(x) = 1, \qquad x \in I_1 \setminus E_1, \quad \mu E_1 = \varepsilon/n; \tag{5}$$

$$\mu\{x \in [0,1]: |\hat{f}_1(x)| > \gamma/2\} < \varepsilon_1/2; \tag{6}$$

$$\|f_1\|_2^2 = \|\hat{f}_1\|_2^2 < 2a^2/\varepsilon n. \tag{7}$$

The unspecified constants $\gamma > 0$ and $\varepsilon_1 > 0$ will be chosen later.

Suppose that the functions $f_1, \dots, f_{\nu-1}$ and the sets $B_1 = [0,1] \subset B_2 \subset \cdots \subset B_{\nu-1}$ have been constructed for $\nu < n$. There is a set B_ν such that

$$\left|\sum_{i=1}^{\nu-1} \hat{f}_i(x)\right| < \frac{\gamma}{2^\nu}, \qquad x \in B_\nu,\ \mu B_\nu < +\infty. \tag{8}$$

It follows from Lemma 3 and (8) that there is a function $f_\nu(x)$ such that

$$f_\nu(x) = 0, \qquad x \notin I_\nu; \tag{9}$$

$$f_\nu(x) = 1, \qquad x \in I_\nu \setminus E_\nu, \quad \mu E_\nu = \varepsilon/n; \tag{10}$$

$$\mu\{x \in B_\nu: |\hat{f}_\nu(x)| > \gamma/2^\nu\} < \varepsilon_1/2^\nu; \tag{11}$$

$$\|\hat{f}_y\|_2^2 < 2a^2/\varepsilon n. \tag{12}$$

Let $f(x) = \sum_1^n f_\nu(x)$. Then (4), (5), (9) and (10) imply 1) and 2) of the lemma. Let us show that n, γ, and ε can be chosen so that $\|\hat{f}\|_q < \varepsilon$.

It is evident that (see (1) and (3))

$$|\hat{f}_\nu(x)| < \frac{M(a)a^2}{\varepsilon n} \quad \text{for all } x \in (0, +\infty) \text{ and } 1 \leqslant \nu < n. \tag{13}$$

Therefore (see (8))

$$\left|\sum_{\nu=1}^{n} f_\nu(x)\right| < \frac{\gamma}{2^n} + \frac{M(a)a^2}{\varepsilon n}, \qquad x \notin B_n. \tag{14}$$

We now write $A_\nu = \{x \in B_\nu: |\hat{f}_\nu(x)| > \gamma/2^\nu\}$. Let $x \in (B_\nu \setminus B_{\nu-1}) \setminus \bigcup_1^n A_i$, $1 < \nu \leqslant n$. Then it follows from (8), (11), and (13) that

$$\begin{aligned} |\hat{f}(x)| = \left|\sum_{i=1}^{n} \hat{f}_i(x)\right| &\leqslant \left|\sum_{i=1}^{\nu-2} \hat{f}_i(x)\right| + |\hat{f}_{\nu-1}(x)| + \sum_{i=\nu}^{n} |f_i(x)| \\ &< \frac{\gamma}{2^{\nu-1}} + \frac{M(a)a^2}{\varepsilon n} + \sum_{i=\nu}^{n} \frac{\gamma}{2^i} < \frac{M(a)a^2}{\varepsilon n} + \gamma. \end{aligned} \tag{15}$$

When $x \in B_1 \setminus \bigcup_1^n A_i$, we have

$$|\hat{f}(x)| \leqslant |\hat{f}_1(x)| + \sum_{i=2}^{n} |\hat{f}_i(x)| < \frac{\gamma}{2} + \sum_{i=2}^{n} \frac{\gamma}{2^i} < \gamma. \tag{16}$$

It follows from (14), (15), and (16) that

$$|\hat{f}(x)| < \frac{M(a)a^2}{\varepsilon n} + \gamma \quad \text{for } x \notin \bigcup_{\nu=1}^{n} A_\nu, \tag{17}$$

and that

$$\mu \bigcup_{\nu=1}^{n} A_\nu < \varepsilon_1. \tag{18}$$

We have (see (13), (17), (18) and (12))

$$\begin{aligned}\int_0^{+\infty} |\hat{f}(x)|^q\,dx &= \int_{\bigcup_{\nu=1}^n A_\nu} |\hat{f}(x)|^q\,dx + \int_{(\bigcup_{\nu+1}^n A_\nu)^c} |\hat{f}(x)|^2 \cdot |\hat{f}(x)|^{q-2}\,dx \\ &\leqslant \left(\frac{M(a)a^2}{\varepsilon}\right)^q \cdot \varepsilon_1 + \left(\frac{M(a)a^2}{\varepsilon n} + \gamma\right)^{q-2} \int_{(\bigcup_{\nu=1}^n A_\nu)^c} |\hat{f}(x)|^2\,dx \\ &\leqslant \left(\frac{M(a)a^2}{\varepsilon}\right)^q \varepsilon_1 + \left(\frac{M(a)a^2}{\varepsilon n} + \gamma\right)^{q-2} \frac{2a^a}{\varepsilon}.\end{aligned} \tag{19}$$

Since $q > 2$, it follows that n can be chosen so large, and γ and ε_1 so small, that the preceding expression is less than ε^q. This completes the proof of the lemma.

PROOF OF THEOREM 2. Let $\{x_k\}_1^\infty$ be dense in $(0, +\infty)$, $p_n \uparrow 2$, $1/p_n + 1/q_n = 1$, and let $\varepsilon_k^n > 0$ be such that $\sum_{n,k=1}^\infty \varepsilon_k^n < \varepsilon$. Then it follows from Lemma 2 that there exist functions $\varphi_k^n(x) \in L_2(0, +\infty)$ such that

1) $\varphi_k^n(x) = 0$, $x \notin (0, x_k)$,
2) $\varphi_k^n(x) = 1$, $x \in (0, x_k) \setminus E_k^n$, $\mu E_k^n \leqslant \varepsilon_k^n$, and
3) $\|\hat{\varphi}_k^n\|_{q_n} < \varepsilon_k^n$.

Let us show that $E = \bigcup_{k,n=1}^\infty E_k^n$ is a set of uniqueness (evidently $\mu E < \varepsilon$). Let $f(x) = 0$ on E and $\hat{f}(x) \in L_p(0, +\infty)$ for some p, $1 \leqslant p < 2$. For sufficiently large n (when $p_n > p$) we have

$$\begin{aligned}\left|\int_0^{x_k} f(x)\,dx\right| &= \left|\int_0^{+\infty} f(x)\varphi_k^n(x)\,dx\right| = \left|\int_0^{+\infty} \hat{f}(x)\hat{\varphi}_k^n(x)\,dx\right| \\ &\leqslant \|\hat{f}\|_{p_n} \cdot \|\varphi_k^n\|_{q_n} < \left(\|\hat{f}\|_p^p + \|\hat{f}\|_2^2\right)^{1/p_n} \varepsilon_k^n.\end{aligned}$$

Since $\lim_{n\to\infty} \varepsilon_k^n = 0$, it follows that

$$\left|\int_0^{x_k} f(x)\,dx\right| = 0.$$

Hence $f(x) = 0$. This completes the proof of Theorem 2.

In conclusion I thank A. A. Talalyan, under whose direction this investigation was carried out.

References

1. Leonardo Colzani, *Sets of uniqueness of l_p for general orthonormal complete systems*, Boll. Un. Mat. Ital. B (5) **16** (1979), 1134–1143.

2. Y. Katznelson, *Sets of uniqueness for some classes of trigonometrical series*, Bull. Amer. Math. Soc. **70** (1964), 722–723.

3. A. A. Talalyan, *The representation of measurable functions by series*, Uspekhi Mat. Nauk **15** (1960), no. 5 (95), 77–141; English transl. in Russian Math. Surveys **15** (1960).

4. ______, *The representation of measurable functions by integrals with kernels of unitary transformations of the space $L_2(0, \infty)$*, Mat. Sb. **53 (95)** (1961), 287–312; English transl. in Amer. Math. Soc. Transl. (2) **44** (1965).

5. G. Alexits, *Konvergenzprobleme der Orthogonalreihen*, Akad. Kiadó, Budapest, 1960; English transl., Pergamon Press, 1961.

Translated by R. P. BOAS

Amer. Math. Soc. Transl.
(2) Vol. **127**, 1986

Singular Integral Equations with Coefficients Having Discontinuities of Semi-Almost-Periodic Type*

UDC 517.983

A. I. SAGINASHVILI

Singular integral equations in various function spaces with various restrictions on the coefficients have been studied by many authors. These investigations are reflected in the monographs of Muskhelishvili [1], Gakhov [2], Khvedelidze [3], and Gohberg and Krupnik [4]. The investigation of singular integral equations with coefficients having discontinuities of the first kind in weighted L_p-spaces was initiated by Khvedelidze (see [3] and [5]), and the Noether theory of these equations assumed a definitive form in the work of Gohberg and Krupnik [4], [6].

Sementsul (see [7] and also [4], Chapter XI) has constructed the theory of one-sided invertibility of singular integral operators with discontinuities of the second kind of almost-periodic type.

In this paper we give criteria for normal solvability, one-sided invertibility and Noether property, and an index formula, for singular integral operators in weighted L_p-spaces in the case where the coefficients of the operator have discontinuities of the second kind of semi-almost-periodic type at points of the contour (in this case the coefficients of the operator have different almost-periodic behavior on different sides of the point of discontinuity). The case that we consider contains as special cases the case of discontinuities of the first kind and the case of discontinuities of almost-periodic type in the coefficients, and combines both these cases as one whole.

The problem is intimately connected with the Riemann boundary-value problem with infinite index, which many authors have studied (see [2], Chapter VI, §49, and the literature cited there).

Singular integral operators with semi-almost-periodic coefficients in the space $L_2(R)$ have been studied by Sarason [8].

1980 *Mathematics Subject Classification*. Primary 45E05; Secondary 30E25.
*Translation of Trudy Tbiliss. Mat. Inst. Razmadze Akad. Nauk Gruzin. SSR **66** (1980), 84–95; MR **82j**: 45006.

The theorems proved here were announced in [9] and [10].

1°. A function of the form

$$f(x) = \sum_{k=1}^{n} c_k e^{i\delta_k x} \qquad (-\infty < x < +\infty),$$

where the c_k are complex and the δ_k are real numbers, is called an *almost-periodic polynomial.*

Let $\mathrm{AP}(R)$ be the closure with respect to the norm of $L_\infty(R)$, $R = (-\infty, +\infty)$, of the set of all almost-periodic polynomials. The set $\mathrm{AP}(R)$ is the subalgebra of $L_\infty(R)$ consisting of all Bohr almost-periodic functions (see [11]).

For every nondegenerate function $f \in \mathrm{AP}(R)$ we define the almost-periodic index

$$\nu(f) = \lim_{l \to +\infty} \frac{1}{2l} [\arg f(x)]_{-l}^{l}. \tag{1}$$

The almost-periodic index $\nu(f)$ does not change for small perturbations of the function f in $\mathrm{AP}(R)$. It is easily seen that $\nu(e^{i\delta x}) = \delta$. Let

$$\lambda(f) = \exp M(g), \tag{2}$$

where

$$g(x) = \ln\left(f(x)e^{-i\nu(f)x}\right), \qquad M(g) = \lim_{T \to +\infty} \frac{1}{2T} \int_{-T}^{T} g(t)\,dt.$$

In this notation the function f admits the representation

$$f(x) = \lambda(f) e^{i\nu(f)x} e^{F(x)}, \tag{3}$$

where $F = g - M(g)$, i.e. $F \in \mathrm{AP}(R)$ and $M(F) = 0$.

Let $\Pi C_\infty(R)$ be the algebra of functions $a(x)$ continuous on R having finite limits $\lim_{x \to \pm\infty} a(x) = a(\pm\infty)$ (generally speaking $a(+\infty) \neq a(-\infty)$).

We denote by $\mathrm{SAP}(R)$ the Banach subalgebra of $L_\infty(R)$ generated by the algebras $\mathrm{AP}(R)$ and $\Pi C_\infty(R)$. If $u_+ \in \Pi C_\infty(R)$, $u_+(+\infty) = 1$, $u_+(-\infty) = 0$ and $u_- = 1 - u_+$ then an arbitrary function from $\mathrm{SAP}(R)$ can be represented uniquely in the form

$$f = u_+ f_+ + u_- f_- + f_0, \tag{4}$$

where $f_+ \in \mathrm{AP}(R)$, $f_0 \in \Pi C_\infty(R)$ and $f_0(\pm\infty) = 0$.

PROPOSITION 1. *The mappings* $f \to f_+$ *and* $f \to f_-$ *are algebraic homomorphisms from* $\mathrm{SAP}(R)$ *into* $\mathrm{AP}(R)$; *moreover,*

$$\inf_{x \in R} |f_\pm(x)| \geqslant \inf_{x \in R} |f(x)|, \tag{5}$$

$$\sup_{x \in R} |f_\pm(x)| \leqslant \sup_{x \in R} |f(x)|. \tag{6}$$

This follows from the definition of an almost-periodic function (see [11], p. 20) and (4).

Functions from the algebra $\mathrm{SAP}(R)$ are called *semi-almost-periodic functions*. They are characterized by the fact that they have different almost-periodic behavior as $x \to +\infty$ and as $x \to -\infty$ (see (4)).

Let $f \in \mathrm{SAP}(R)$ and $\inf_{x \in R}|f(x)| > 0$; then, by (5), $\inf_{x \in R}|f_{\pm}(x)| > 0$. If now the functions f_+ and f_- are represented in the form (3), then we obtain from (4) a representation of f in the form

$$f(x) = \lambda_+(f)u_+(x)e^{i\nu_+(f)}e^{F_+(x)} + \lambda_-(f)u_-(x)e^{i\nu_-(f)x}e^{F_-(x)} + f_0, \quad (7)$$

where $\lambda_{\pm}(f) = \lambda(f_{\pm})$, $\nu_{\pm}(f) = \nu(f_{\pm})$ (see (1) and (2)), while $F_{\pm} \in \mathrm{AP}(R)$ and $M(F_+) = M(F_-) = 0$; we call $\nu_+(f)$ and $\nu_-(f)$ the *almost-periodic indices* of the function f.

Let Γ be a simple smooth closed contour, $t_0 \in \Gamma$ and $\omega\colon \Gamma \to R$ a diffeomorphism, $\omega' \in C(\Gamma)$, $\omega'(t) \neq 0$, for which

$$\lim_{t \to t_0 \pm 0} \omega(t) = \mp\infty.$$

DEFINITION. We call the point $t_0 \in \Gamma$ a *point of discontinuity of semi-almost-periodic type* of the function $a \in L_\infty(\Gamma)$ if there exists a function $f \in \mathrm{SAP}(R)$ such that

$$\lim_{t \to t_0} (a(t) - f(\omega(t))) = 0.$$

In what follows, for simplicity we consider functions that have discontinuities of semi-almost-periodic type only at one point $t_0 \in \Gamma$ and are continuous at all remaining points of the contour, although all our results carry over without special effort to the case of finitely many such discontinuities; we denote the set of such functions by $\mathrm{SAP}(\Gamma, t_0, \omega)$.

Obviously $\mathrm{SAP}(\Gamma, t_0, \omega)$ is a Banach algebra; if $a \in \mathrm{SAP}(\Gamma, t_0, \omega)$ then $a \circ \omega^{-1} \in \mathrm{SAP}(R)$; when $\inf_{t \in \Gamma}|a(t)| > 0$ we introduce the notation

$$\nu_{\pm}(a) = \nu_{\pm}(a \circ \omega^{-1}) \quad \text{and} \quad \lambda_{\pm}(a) = \lambda_{\pm}(a \circ \omega^{-1});$$

it is natural to call the numbers $\nu_+(a)$ and $\nu_-(a)$ the *semi-almost-periodic indices* of the function a.

2°. In this section we present some known results about singular integral equations that will be needed later.

As usual we denote by $L_p(\Gamma, \rho)$, $1 < p < \infty$, the Banach space of all measurable functions, pth power integrable, with weight

$$\rho(t) = \prod_{k=1}^{n} |t - \tau_k|^{\beta_k}, \qquad -1 < \beta_k < p - 1. \quad (8)$$

Let S_Γ be the operator of singular integration

$$S_\Gamma\varphi(t) = \frac{1}{\pi i}\int_\Gamma \frac{\varphi(\tau)}{\tau - t}\,d\tau,$$

and P_Γ and Q_Γ singular projections

$$P_\Gamma = \tfrac{1}{2}(I + S_\Gamma), \qquad Q_\Gamma = \tfrac{1}{2}(I - S_\Gamma).$$

A bounded linear operator A acting in the Banach space X is called a Φ_+ (Φ_-)-*operator* if the operator A is normally solvable and the dimension of its kernel is finite (the dimension of the kernel of the adjoint operator in the dual space is finite). The operator A is called *Noetherian* if it is simultaneously a Φ_+- and a Φ_--operator. The index of a Noetherian operator A, equal to the difference $\dim \operatorname{Ker} A - \dim \operatorname{Ker} A^*$, is denoted by $\operatorname{Ind} A$.

The following two theorems are due to Simonenko (see [12] and also [4], Chapter VII).

THEOREM 1. *Let $a \in L_\infty(\Gamma)$, $a(t) \not\equiv 0$ and $A = aP_\Gamma + Q_\Gamma$, then at least one of the numbers* $\dim \operatorname{Ker} A$ *and* $\dim \operatorname{Ker} A^*$ *is equal to zero in the space $L_p(\Gamma, \rho)$.*

THEOREM 2. *Let $a \in L_\infty(\Gamma)$ and $a(t) \not\equiv 0$. If the operator $aP_\Gamma + Q_\Gamma$ is normally solvable in the space $L_p(\Gamma, \rho)$ then* $\operatorname{ess\,inf}_{t\in\Gamma}|a(t)| > 0$.

Often we shall make use of the following theorem (see [13], and also [14]).

THEOREM 3. *The operator $aP_\Gamma + Q_\Gamma$ in the space $L_p(\Gamma, \rho)$ where ρ is defined by* (1.8), *and the operator $ahP_\Gamma + Q_\Gamma$ in the space $L_p(\Gamma)$, where $h(t) = \prod_1^n h_k(t)$ and h_k is a branch of the function $t^{-\beta_k/p}$, continuous on Γ/τ_k, are connected by the equality*

$$B_1(aP_\Gamma + Q_\Gamma)B_2 = ahP_\Gamma + Q_\Gamma,$$

where B_1 and B_2 are invertible operators.

We associate with every function $a(t)$, continuous on $\Gamma \setminus t_0$ and having a discontinuity of the first kind at the point t_0, the function $a_{p,\rho}(t, x)$, $t \in \Gamma$, $x \in \overline{R}$ in the space $L_p(\Gamma, \rho)$:

$$a_{p,\rho}(t, x) = \tfrac{1}{2}(a(t-0) + a(t+0)) - \tfrac{1}{2}(a(t-0) - a(t+0)) \coth \pi(i\alpha_t + x), \tag{9}$$

where $\alpha_t = (1 + \beta_t)/p$ and β_t is the exponent of the weight at the point t.

If $\inf_{t\in\Gamma, x\in R}|a_{p,\rho}(t, x)| > 0$, then by equality

$$\operatorname{ind}_{p,\rho} a = \frac{1}{2\pi}\left[\arg a(t)\right]_{t_0+0}^{t_0-0} + \frac{1}{2\pi}\left[\arg a_{p,\rho}(t_0, x)\right]_{-\infty}^{+\infty}$$

defines the integer (p, ρ)-index of the function a; here $\arg a(t)$ and $\arg a_{p,\rho}(t_0, x)$ are taken continuous on $\Gamma \setminus t_0$ and R, respectively.

Let the operator A have the form

$$A = \sum_{k=1}^{m} A_{k_1} \cdots A_{k_s}, \tag{10}$$

where $A_{k_j} = a_{k_j}P_\Gamma + Q_\Gamma$ and the functions a_{k_j} are continuous everywhere except at the points $t_k \in \Gamma$ at which they have discontinuities of the first kind; by the *symbol* of the operator A in $L_p(\Gamma, \rho)$ we mean the function

$$A_{p,\rho}(t, x) = \sum_{k=1}^{n} (a_{k_1})_{p,\rho}(t, x) \cdots (a_{k_s})_{p,\rho}(t, x).$$

In the case $\inf|A_{p,\rho}(t, x)| > 0$ the index $\operatorname{ind}_{p,\rho} A$ is defined in the natural way.

THEOREM 4 (see [6]). *In order for the operator A of the form* (10) *to be a Φ_+- or Φ_--operator in $L_p(\Gamma, \rho)$ it is necessary and sufficient that its symbol be nondegenerate*: $\inf|A_{p,\rho}(t,x)| > 0$. *If this condition is satisfied, then the operator is Noetherian and* $\operatorname{Ind} A = -\operatorname{ind}_{p,\rho} A$.

We often use the following lemma of Sarason (see [15]).

LEMMA 1. *Let $u \in \Pi C_\infty(R)$ and $h \in H_\infty$ be such that $h(iy) \to 0$ as $y \to +\infty$; then $u \cdot h \in H_\infty + C$, where C is the set of functions continuous on $\mathring{\mathbf{R}}$, the one-point compactification of R.*

3°. The following two theorems are the main results of this paper.

THEOREM I. *Let $a \in \mathrm{SAP}(\Gamma, t_0, \omega)$ and $a(t) \not\equiv 0$. For normal solvability of the operator $aP_\Gamma + Q_\Gamma$ in the space $L_p(\Gamma, \rho)$ it is necessary and sufficient that the following conditions hold*:

1. $\inf_{t\in\Gamma}|a(t)| > 0$.
2. $\nu_+(a) \cdot \nu_-(a) \geqslant 0$.
3. *If* $\nu_+(a) = \nu_-(a) = 0$ *then*

$$\frac{1+\beta_0}{p} - 1 < \frac{1}{2\pi}\frac{\lambda_+(a)}{\lambda_-(a)} < \frac{1+\beta_0}{p}, \tag{11}$$

where β_0 is the exponent of the weight ρ at the point t_0.

Let $a \in \mathrm{SAP}(\Gamma, t_0, \omega)$, $\inf|a(t)| > 0$ and $\nu_+(a) = \nu_-(a) = 0$; then the function

$$\tilde{a}(t) = a(t)\exp[-u_+(\omega(t))F_+(\omega(t)) - u_-(\omega(t))F_-(\omega(t))], \tag{12}$$

where F_+ and F_- are functions from a decomposition of the form (7) for the function $a \circ \omega^{-1}$, has at the point t_0 a discontinuity of the first kind, $a(t_0 \pm 0) = \lambda_\mp$, and condition (11) is equivalent to the condition $\tilde{a}_{p,\rho}(t,x) \neq 0$, $t \in \Gamma$, $x \in R$.

THEOREM II. *If the operator $aP_\Gamma + Q_\Gamma$ is normally solvable, then the following assertions are true*:

1) *For $\nu_+(a) = \nu_-(a) = 0$ the operator $aP_\Gamma + Q_\Gamma$ is Noetherian in the space $L_p(\Gamma, \rho)$, and*

$$\operatorname{Ind}(aP_\Gamma + Q_\Gamma) = -\operatorname{ind}_{p,\rho}\tilde{a},$$

where $\tilde{a}$ is defined by (12).

2) *For $\nu_+(a) + \nu_-(a) > 0$ the operator $aP_\Gamma + Q_\Gamma$ is invertible from the left, and* $\dim\operatorname{Coker}(aP_\Gamma + Q_\Gamma) = +\infty$.

3) *For $\nu_+(a) + \nu_-(a) < 0$ the operator $aP_\Gamma + Q_\Gamma$ is invertible from the right, and* $\dim\operatorname{Ker}(aP_\Gamma + Q_\Gamma) = +\infty$.

4°. In this section we give some lemmas upon which the proof of Theorems I and II relies.

LEMMA 2. *Let the functions F_+ and F_- belong to* AP(R), *with* $M(F_+) = M(F_-) = 0$. *The operators* $fP_R + Q_R$ *and* $\tilde{f}P_R + Q_R$, *where* $f, \tilde{f} \in L_\infty(R)$ *are related by the equality*

$$f = \tilde{f}\exp(u_+F_+ + u_-F_-), \tag{13}$$

are Noether equivalent[1] *in* $L_p(R, \rho_0)$, $\rho_0(x) = |x - i|^\beta \prod_1^n |x - x_k|^{\beta_k}$, $-1 < \beta_k < p - 1$, $-1 < \beta + \sum_1^n \beta_k < p - 1$, *and their indices coincide.*

PROOF. Let $fP_R + Q_R$ be a Φ_+-operator; since the set of trigonometric polynomials is dense in AP(R), the functions F_+ and F_- from (13) can be replaced by the trigonometrical polynomials $r_\pm(x) = \sum_0^m c_j^\pm \exp(i\lambda_j^\pm x)$ such that the operator $\tilde{f}e^{u_+r_+ + u_-r_-}P_R + Q_R$ is again a Φ_+-operator and the index of the operator is unchanged; moreover, in view of the condition $M(F_\pm) = 0$ one can assume that $c_j^\pm = 0$ for $\lambda_j^\pm = 0$.

Let $r_+ = q_+ + s_+$ and $r_- = q_- + s_-$ where

$$q_\pm(x) = \sum_{\lambda_j^\pm > 0} c_j^\pm \exp(i\lambda_i^\pm x), \qquad s_\pm(x) = \sum_{\lambda_j^\pm < 0} c_j^\pm \exp(i\lambda_j^\pm x)$$

according to Lemma 1. The functions $u_+q_+ + u_-q_-$ and $\overline{u_+s_+ + u_-s_-}$ belong to the class $H_\infty + C$; therefore

$$\tilde{f}e^{u_+r_+ + u_-r_-}P_R + Q_R = (e^{h_1}P_R + Q_R)(\tilde{f}P_R + Q_R)(e^{h_2}P_R + Q_R) + T, \tag{14}$$

where $h_1 = u_+s_+ + u_-s_-$, $h_2 = u_+q_+ + u_-q_-$, and T is a compact operator in $L_p(R, \rho_0)$. It is easily seen that the operators $e^{h_1}P_R + Q_R$ and $e^{h_2}P_R + Q_R$ are invertible in $L_p(R, \rho_0)$; hence, by, (14), $\tilde{f}P_R + Q_R$ is a Φ_+-operator. The remaining assertions of the lemma can be proved by analogous reasoning.

Let the functions f and g be defined by

$$f(x) = \lambda_+u_+(x)e^{i\nu_+x} + \lambda_-u_-(x)e^{i\nu_-x} + f_0(x), \tag{15}$$

$$g(x) = \chi_+(x)e^{i\nu x_+} + \chi_-(x)e^{i\nu x_-}, \tag{16}$$

where $\lambda_\pm$ are complex numbers, $\nu_\pm$ real numbers, χ_+ the characteristic function of the positive semiaxis, $\chi_- = 1 - \chi_+$, and $f_0 \in C$, $f_0(\infty) = 0$.

LEMMA 3. *If* $\inf_{x\in R}|f(x)| > 0$ *and* ν_+ *or* ν_- *is different from zero, then the operators* $fP_R + Q_R$ *and* $gP_R + Q_R$, *where* f *and* g *are defined by* (15) *and* (16), *are Noether equivalent.*

PROOF. Without loss of generality one can assume that $\nu_+ \neq 0$ and also $|\lambda_+| = \lambda_- = 1$. Let $\lambda_+ = e^{i\mu}$; then

$$f(x) = u_+(x)e^{i(\nu_+x + \mu)} + u_-(x)e^{i\nu_-x} + f_0(x).$$

Let the function $x = v(y)$ be defined by

$$v(y) = \begin{cases} y - \mu/\nu_+ & \text{for } y > N, \\ v_0(y) & \text{for } -N \leqslant y \leqslant N, \\ y & \text{for } y < -N, \end{cases}$$

(1) We say that two operators are *Noether equivalent* if they are simultaneously Noetherian, have left or right regularizers and are Φ_+- or Φ_--operators.

where $N > |\mu/2\nu_+|$ and the function v_0 is taken so that v is a diffeomorphism of R onto R, $v' \in C$. Let $V\varphi(x) = \varphi(v(x))$; then

$$V(fP_R + Q_R)V^{-1} = (f \circ v)P_R + Q_R + T, \tag{17}$$

where T is a compact operator, but, as is easily seen, $f \circ v = g \cdot c$, where $c \in C$; then

$$\begin{aligned} f \circ vP_R + Q_R &= (gP_R + Q_R)(cP_R + Q_R) + T_1 \\ &= (cP_R + Q_R)(gP_R + Q_R) + T_2, \end{aligned}$$

where T_1 and T_2 are compact operators; now from (17) we obtain

$$\begin{aligned} V(aP_R + Q_R)V^{-1} &= (gP_R + Q_R)(cP_R + Q_R) + T_3 \\ &= (cP_R + Q_R)(gP_R + Q_R) + T_4, \end{aligned}$$

where T_3 and T_4 are compact operators; taking into account that $cP_R + Q_R$ is a Noetherian operator in $L_p(R, \rho)$ we now obtain the statement of the lemma.

LEMMA 4. *Let $f(t) = \chi_+(x)e^{i\nu_+ t} + \chi_-(x)e^{i\nu_- t}$, where ν_+ and ν_- are real numbers. If at least one of the numbers ν_+ and ν_- is strictly positive, then the operator $fP_R + Q_R$ has zero kernel and an infinite-dimensional algebraic complement to the image in the space $L_p(R)$, $1 < p \leqslant 2$.*

PROOF. Let F be the Fourier transformation

$$F\varphi(t) = \int_{-\infty}^{+\infty} e^{ixt}\varphi(x)\,dx,$$

and F^{-1} the transformation inverse to it. For the function $\varphi \in L_p(R)$, $1 < p \leqslant 2$, the following equalities are valid (see, for instance, [16], Corollary 1.36):

$$F^{\pm 1}P_R\varphi = \chi_\mp F^{\pm 1}\varphi, \qquad F^{\pm 1}Q_R\varphi = \chi_\pm F^{\pm 1}\varphi. \tag{18}$$

We assume $\nu_+ > 0$ and let $\varphi \in L_p(R)$, $1 < p \leqslant 2$, be such that $fP_R\varphi + Q_R\varphi = 0$. Applying the inverse Fourier transformation to this equality and using (18), we obtain

$$\tau_{\nu+}Q_R\chi_+F^{-1}\varphi + \tau_{\nu-}P_R\chi_-F^{-1}\varphi = -\chi_-F^{-1}\varphi. \tag{19}$$

Because $\chi_+F^{-1}\varphi(x) = 0$ for $x < 0$, we have

$$\tau_{\nu+}Q_R\chi_+F^{-1}\varphi(x) = -\tau_{\nu+}P_R\chi_+F^{-1}\varphi(x), \qquad 0 < x < \nu_+; \tag{20}$$

substituting (20) into (19), we obtain

$$\tau_{\nu+}P_R\chi_+F^{-1}\varphi(x) = \tau_{\nu-}P_R\chi_+F^{-1}\varphi(x), \qquad 0 < x < \nu_+. \tag{21}$$

From the uniqueness theorem for analytic functions (see [17], Chapter IV, §2.5) it follows that (21) is valid on all of R; substituting (21) into (19), we obtain

$$\tau_{\nu+}\left(P_R\chi_+F^{-1}\varphi(x) + Q_R\chi_+F^{-1}\varphi(x)\right) = -\chi_-F^{-1}\varphi(x)$$

or

$$\tau_{\nu+}\chi_+F^{-1}\varphi(x) = -\chi_-F^{-1}\varphi(x);$$

wherefore $F^{-1}\varphi(x) = 0$ and $\varphi(x) = 0$ for $x \in R$.

Let

$$\varphi_\varepsilon(x) = (e^{i\varepsilon x} - 1)/ix,$$

where $0 < \varepsilon < \nu_+$. We prove that the image of the operator $fP_R + Q_R$ in the space $L_p(R)$, $1 < p \leqslant 2$, does not contain the function φ_ε. We assume that there exists $\varphi \in L_p(R)$, $1 < p \leqslant 2$, such that $fP_R\varphi + Q_R\varphi = \varphi_\varepsilon$. Acting on this equality with the operator F^{-1} and taking into account that $F^{-1}\varphi_\varepsilon = \chi_{(0,\varepsilon)}$, where $\chi_{(0,\varepsilon)}$ is the characteristic function of the interval $(0, \varepsilon)$, we obtain, by the same method as before,

$$\tau_{\nu+}Q_R\chi_+F^{-1}\varphi + \tau_{\nu-}P_R\chi_+F^{-1}\varphi = -\chi_-F^{-1}\varphi + \chi_{(0,\varepsilon)}.$$

The function on the left side of this equality can be analytically continued in the upper half plane to the interval $(0, \nu_+)$; but the right side on this interval is equal to $\chi_{(0,\varepsilon)}$, which cannot have such a property. This contradiction proves that the φ_ε do not belong to the image of the operator $fP_R + Q_R$; but since $\{\varphi_\varepsilon\}_{\varepsilon \in (0,\nu_+)}$ is a linearly independent system in $L_p(R)$, the image of the operator $fP_R + Q_R$ has an infinite-dimensional algebraic complement.

Analogous reasoning can be carried out for the case $\nu_- > 0$. The lemma is proved.

LEMMA 5. *Let $f \in \mathrm{SAP}(R)$ have the form*

$$f(x) = \lambda_+u_+(x)e^{i\nu x} + \lambda_-u_-(x) + f_0(x),$$

where λ_+ and λ_- are complex numbers, $\nu > 0$ and $f_0 \in C$ and $f(\pm\infty) = 0$. If $\inf_{x\in R}|f(x)| > 0$ *then the operator $fP_R + Q_R$ is invertible from the left in $L_p(R, \rho_0)$, and* $\dim \operatorname{Coker}(fP_R + Q_R) = +\infty$.

PROOF. First we prove that the operator $fP_R + Q_R$ admits a left regularization. Since

$$fP_R + Q_R = (P_RfP_R + Q_R)(I + Q_RfP_R),$$

and $I + Q_RfP_R$ is invertible, $(I + Q_RfP_R)^{-1} = (I - Q_RfP_R)$, it is sufficient to prove that the operator $P_RfP_R + Q_R$ admits a left regularization. Let

$$f_1(x) = \frac{4}{\lambda_+}u_+(x)e^{-i\nu x} + \frac{1}{\lambda_-}u_-(x) + f_{1_0},$$

where $f_{1_0} \in C$, $f_{1_0}(\infty) = 0$ and $\inf_{x\in R}|f_1(x)| > 0$. We consider the operator

$$A = (P_Rf_1P_R + Q_R)(P_RfP_R + Q_R); \tag{22}$$

with the help of Lemma 1 we easily obtain

$$A = P_R\left(4u_+^2 + f_{2_0}\right)P_R + (P_Ru_-P_R)^2 + Q_R + T,$$

where $f_{2_0} \in C$, $f_{2_0}(\infty) = 0$ and T is a compact operator.

It is easy to check that the symbol of the operator A in $L_p(R, \rho_0)$ can be calculated by the formula

$$A_{p,\rho_0}(t, x) = \begin{cases} f_1(t)\cdot f(t) & \text{for } t \in R, \\ \dfrac{1}{2}\left(\coth\pi\left(i\dfrac{1+\beta_0}{p} + x\right) - 3\right)^2 & \text{for } |t| = \infty,\ x \in \overline{R}, \end{cases}$$

where $\beta_0 = p - 2 - \beta - \Sigma_1^n \beta_k$ (here we use formulas (9) and (10) for the case of the real axis).

Since $\inf|A_{p,\rho_0}(t,x)| > 0$, it follows from Theorem 4 that A admits a left regularization in $L_p(R,\rho_0)$, and from (22) we obtain that also $P_R f P_R + Q_R$ admits a left regularization in $L_p(R,\rho_0)$.

It remains to prove that $\dim \operatorname{Coker}(fP_R + Q_R) = +\infty$, since according to Theorem 1 this implies also the invertibility of the operator $fP_R + Q_R$ from the left side.

First let $1 < p \leqslant 2$. The operator $fP_R + Q_R$ is Noether equivalent to the operator $gP_R + Q_R$, where $g(x) = \chi_+(x)e^{i\nu x} + \chi_-(x)$ (see Lemma 3); it follows, therefore, from Lemma 4 that $\dim \operatorname{Coker}(fP_R + Q_R) = +\infty$ in $L_p(R)$. We now consider the case $L_p(R,\rho_0)$, where $\rho_0(x) = |x - i|^\beta \Pi_1^n |x_k - x|^{\beta_k}$, $-1 < \beta_k < p - 1$, $-1 < \beta + \Sigma_1^n \beta_k < p - 1$. Let $h = h_0 \Pi_1^n h_k$, where h_k, $k = 1,\ldots,n$, is a branch of the function $((x-i)/(x+i))^{-\beta_k/p}$ continuous on $\mathring{\mathbf{R}} \setminus x_k$ and h_0 is a branch of $((x-i)/(x+i))^{-\beta_0/p}$ continuous on R, where $\beta_0 = p - 2 - \beta - \Sigma_1^n \beta_k$. It is not hard to see that the operators $fhP_R + Q_R$ and $fh_0 P_R + Q_R$ are Noether equivalent in $L_p(R)$. But since fh_0 satisfies the conditions of the lemma, it follows from what has already been proved that $\dim \operatorname{Coker}(fh_0 P_R + Q_R) = +\infty$ and, hence, $\dim \operatorname{Coker}(fhP_R + Q_R) = +\infty$ in $L_p(R)$. Applying Theorem 3 to the case $\Gamma = R$, we obtain that $\dim \operatorname{Coker}(fP_R + Q_R) = +\infty$ in $L_p(R,\rho_0)$.

Now let $p > 2$. According to what has already been proved the operator adjoint to $fP_R + Q_R$ (one can take $\bar{f}P_R + Q_R$ for it) has an infinite-dimensional kernel in $L_p(R,\rho_p)$, where $\rho_p = |x - i|^{p-2}$. But $L_p(R,\rho_p) \subset L_q(R,\rho_q)$, where $q = p/(p-1)$, and so the operator $\bar{f}P_R + Q_R$ has an infinite-dimensional kernel in the space $L_q(R,\rho_q)$ dual to $L_p(R,\rho_p)$. This means that $\dim \operatorname{Coker}(fP_R + Q_R) = +\infty$. The passage to an arbitrary weight ρ_0 occurs in the same way as for $1 < p \leqslant 2$. Lemma 5 is completely proved.

5°. Proof of Theorems I and II. The necessity of condition 1 of Theorem I follows from Theorem 2. The rest of the theorem we prove only for the case of the space $L_p(R,\rho_0)$, $1 < p < \infty$, since the passage to $L_p(\Gamma,\rho)$ is standard (see [4], Chapter XI, §3, proof of Theorem 3.1, or [18]).

Let $f \in \mathrm{SAP}(R)$ and $\inf_{t\in R}|f(t)| > 0$. According to (7), f can be represented in the form

$$f(x) = \lambda_+(f)u_+(x)e^{i\nu_+(f)x}e^{F_+(x)} + \lambda_-(f)u_-(x)e^{i\nu_-(f)x}e^{F_-(x)} + f_0(x).$$

According to Lemma 2 the operators $fP_R + Q_R$ and $\tilde{f}P_R + Q_R$ are Noether equivalent and in case of the Noether property their indices coincide. Therefore, it suffices to conduct the proof for the operator $\tilde{f}P_R + Q_R$.

It is easily seen that $\tilde{f}$ has the form

$$\tilde{f}(x) = \lambda_+(f)u_+(x)e^{i\nu_+(f)x} + \lambda_-(f)u_-(x)e^{i\nu_-(f)x} + \tilde{f}_0(x),$$

where $\tilde{f}_0 \in C$, $\tilde{f}_0(\infty) = 0$.

Let $\nu_+(f) = \nu_-(f) = 0$; then $\tilde{f} \in \Pi C_\infty(R)$, $\tilde{f}(\pm\infty) = \lambda_\pm(f)$ and, according to Theorem 4, condition (11) with $\beta_0 = p - 2 - \beta - \Sigma_1^n \beta_k$ is necessary and sufficient for the normal solvability of the operator $\tilde{f}P_R + Q_R$ in $L_p(R, \rho_0)$. If this condition is satisfied the operator is Noetherian and

$$\operatorname{Ind}(\tilde{f}P_R + Q_R) = -\operatorname{ind}_{p,\rho_0}\tilde{f}.$$

In the case $\nu_+(f) > 0$ and $\nu_-(f) = 0$, according to Lemma 5 the operator $\tilde{f}P_R + Q_R$ is invertible from the left and $\dim \operatorname{Coker}(\tilde{f}P_R + Q_R) = +\infty$. The proof in the case $\nu_+(f) = 0$ and $\nu_-(f) > 0$ is analogous.

If $\nu_+(f) > 0$ and $\nu_-(f) > 0$, then it follows from Lemma 1 that $\tilde{f} \in H_\infty + C$; therefore the operator $\tilde{f}P_R + Q_R$ has the left regularizer $1/\tilde{f}P_R + Q_R$. The equality $\dim \operatorname{Coker}(\tilde{f}P_R + Q_R) = +\infty$ is proved in exactly the same way as in the case $\nu_+(f) > 0$, $\nu_-(f) = 0$.

Now let $\nu_+(f) \cdot \nu_-(f) < 0$. We prove that in this case the operator $\tilde{f}P_R + Q_R$ is not normally solvable in $L_p(R, \rho_0)$. As follows from Theorem 1, it is sufficient to prove that $\tilde{f}P_R + Q_R$ is neither a Φ_+- nor a Φ_--operator. We prove this for $1 < p \leqslant 2$ (the case $p > 2$ is obtained by passage to the adjoint operator). We assume $\tilde{f}P_R + Q_R$ to be a Φ_--operator in $L_p(R, \rho_0)$. Applying Theorem 3 and Lemma 3 in the same manner as for the proof of Lemma 5, we obtain that $\tilde{f}P_R + Q_R$ is also a Φ_--operator in $L_p(R)$. Then it follows from Lemma 3 that $gP_R + Q_R$ is also a Φ_--operator in $L_p(R)$, where

$$g(x) = \chi_+(x)e^{i\nu_+(f)x} + \chi_-(x)e^{i\nu_-(f)x}.$$

But according to Lemma 4 $\dim \operatorname{Coker}(gP_R + Q_R) = +\infty$. This contradiction proves that $\tilde{f}P_R + Q_R$ is not a Φ_--operator in $L_p(R, \rho_0)$. We assume now that $\tilde{f}P_R + Q_R$ is a Φ_+-operator in $L_p(R, \rho_0)$; then by the same reasoning as before we obtain that $\tilde{f}P_R + Q_R$ is a Φ_+-operator in $L_p(R, \rho_p)$, and by passage to the adjoint operator we have that $\bar{\tilde{f}}P_R + Q_R$ is a Φ_--operator in $L_q(R, \rho_q)$ and $\dim \operatorname{Ker}(\bar{\tilde{f}}P_R + Q_R) = +\infty$. According to Lemma 3 the operator $\bar{g}P_R + Q_R$ also possesses this property. But since $L_q(R, \rho_q) \subset L_2(R)$, according to Lemma 4 the operator $\bar{g}P_R + Q_R$ has zero kernel. This contradiction proves that $\tilde{f}P_R + Q_R$ is not a Φ_+-operator in $L_p(R, \rho_0)$.

All the remaining cases reduce to those already considered by passage to the adjoint operator. Theorems I and II are completely proved.

References

1. N. I. Muskhelishvili, *Singular integral equations*, 3rd ed., "Nauka", Moscow, 1968; English transl. of 1st ed., Noordhoff, 1953.

2. F. D. Gakhov, *Boundary value problems*, 3rd ed., "Nauka", Moscow, 1977; English transl. of 2nd ed., Pergamon Press, Oxford, and Addison-Wesley, Reading, Mass., 1966.

3. B. V. Khvedelidze, *The method of Cauchy-type integrals in discontinuous boundary-value problems of the theory of holomorphic functions of a complex variable*, Itogi Nauki: Sovremennye Problemy Mat., vol. 7, VINITI, Moscow, 1975, pp. 5–162; English transl. in J. Soviet Math. **7** (1977), no. 5.

4. I. Ts. Gokhberg [Israel Gohberg] and N. Ya. Krupnik, *Introduction to the theory of one-dimensional singular integral operators*, "Shtiintsa", Kishinev, 1973; German transl., Birkhäuser, 1978.

5. B. V. Khvedelidze, *Linear discontinuous boundary value problems in the theory of functions, singular integral equations and some of their applications*, Akad. Nauk Gruzin. SSR Trudy Tbiliss. Mat. Inst. Razmadze **23** (1956), 3–158. (Russian)

6. I. Ts. Gokhberg [Israel Gohberg] and N. Ya. Krupnik, *Singular integral operators with piecewise continuous coefficients and their symbols*, Izv. Akad. Nauk SSSR Ser. Mat. **35** (1971), 940–964; English transl. in Math. USSR Izv. **5** (1971).

7. A. A. Sementsul, *Singular integral equations with coefficients that have discontinuities of almost periodic type*, Mat. Issled. **6** (1971), no. 3(21), 92–114. (Russian)

8. Donald Sarason, *Toeplitz operators with semi-almost periodic symbols*, Duke Math. J. **44** (1977), 357–364.

9. A. I. Saginashvili, *Singular integral operators with coefficients having semi-almost-periodic discontinuities*, Soobshch. Akad. Nauk Gruzin. SSR **94** (1979), 289–291. (Russian)

10. ______, *Singular integral operators with coefficients having semi-almost-periodic discontinuities*, SSR **95** (1979), 541–543. (Russian)

11. B. M. Levitan, *Almost-periodic functions*, GITTL, Moscow, 1953. (Russian)

12. I. B. Simonenko, *Some general questions in the theory of the Riemann boundary value problem*, Izv. Akad. Nauk SSSR Ser. Mat. **32** (1968), 1138–1146; English transl. in Math. USSR Izv. **2** (1968).

13. N. Ya. Krupnik and V. I. Nyaga, *Singular operators in L_p spaces with a weight*, Mat. Issled. **9** (1974), no. 3(33), 206–209. (Russian)

14. V. A. Paatashvili, *On a discontinuous problem of linear conjugacy*, Soobshch. Akad. Nauk Gruzin. SSR **36** (1964), 539–540. (Russian)

15. Donald Sarason, *Approximation of piecewise continuous functions by quotients of bounded analytic functions*, Canad. J. Math. **24** (1972), 642–657.

16. R. V. Duduchava, *Integral equations in convolution with discontinuous presymbols, singular integral equations with fixed singularities, and their applications to some problems of mechanics*, Akad. Nauk Gruzin. SSR Trudy Tbiliss. Mat. Inst. Razmadze **60** (1979); English transl., Teubner, Leipzig, 1979.

17. I. I. Privalov, *Boundary properties of analytic functions*, 2nd ed., GITTL, Moscow, 1950; German transl., VEB Deutscher Verlag Wiss., Berlin, 1956.

18. S. M. Grudskiĭ and V. B. Dybin, *The Riemann boundary value problem with discontinuities of almost periodic type in its coefficient*, Dokl. Akad. Nauk SSSR **237** (1977), 21–24; English transl. in Soviet Math. Dokl. **18** (1974).

Translated by E. LÜNEBURG

Amer. Math. Soc. Transl.
(2) Vol. **127**, 1986

Application of Interpolation Methods to Estimates of the Spectrum of Integral Operators*

M. SH. BIRMAN AND M. Z. SOLOMYAK

The object of this report is to describe the present status of the question of how singular numbers (s-numbers) of an integral operator behave, depending on the smoothness of its kernel. Because of the requirements of the applications, we have to study rather extended classes of equations: the multidimensional case; unbounded domains; kernels which are not square integrable; operators in weight spaces; various special types of kernels (especially kernels of Green's function type). We likewise study multipliers in classes of integral kernels.

Obtaining estimates under such broad conditions required a new method. It is based on a special method of piecewise-polynomial approximation (see [1] and [2]) for functions in the Sobolev space W_p^α. As a result there arises a sufficiently flexible machinery for the approximation of a kernel by degenerate kernels, leading to estimates for the s-numbers. It is essential that the estimates are uniform with respect to a suitable class of measures (weights). Another characteristic feature of the method is the a posteriori sharpening and extension of the domain of applicability of the estimates by means of linear and bilinear interpolation techniques. Here it is not possible to restrict oneself to interpolation of normed classes of operators only, but one is led to consider quasi-normed (q.n.) classes too.

§1. Preliminaries and initial results

Let $\mathfrak{H}_1$ and $\mathfrak{H}_2$ be separable Hilbert spaces, $\mathbf{T} \in \mathfrak{S}_\infty = \mathfrak{S}_\infty(\mathfrak{H}_1, \mathfrak{H}_2)$ (a compact operator from $\mathfrak{H}_1$ into $\mathfrak{H}_2$) and $s_n = s_n(\mathbf{T}) = \lambda_n(\sqrt{\mathbf{T}^*\mathbf{T}})$ the s-numbers of $\mathbf{T}$. By estimates of s-numbers we usually intend the inequalities

$$s_n(\mathbf{T}) \leqslant Cn^{-\omega}, \qquad 0 < \omega < \infty; \tag{1}$$

$$\sum_n s_n^p(\mathbf{T}) < C, \qquad 0 < p < \infty. \tag{2}$$

1980 *Mathematics Subject Classification*. Primary 47G05, 47A10; Secondary 46E35.

*Translation of Operator Theory in Function Spaces (Proc. School, Novosibirsk, 1975; G. P. Akilov, editor), "Nauka", Novosibirsk, 1977, pp. 42–70; MR **80b**: 47063.

A natural generalization leads to the classes $\mathfrak{S}_{p,q} \subset \mathfrak{S}_\infty$: $\mathbf{T} \in \mathfrak{S}_{p,q}$ $(0 < p < \infty,$ $0 < q \leqslant \infty)$ if the functional $N(\mathbf{T}|\mathfrak{S}_{p,q})$ is finite, where

$$N^q(\mathbf{T}|\mathfrak{S}_{p,q}) = \sum_n n^{qp^{-1}-1} s_n^q(\mathbf{T}) \qquad (0 < q < \infty);$$

$$N(\mathbf{T}|\mathfrak{S}_{p,\infty}) = \sup_n n^{p^{-1}} s_n(\mathbf{T}). \tag{3}$$

The relation $\mathbf{T} \in \mathfrak{S}_{p,\infty}$ obviously corresponds to the estimate (1), with $p = 1$, and $\mathbf{T} \in \mathfrak{S}_p \overset{\text{def}}{=} \mathfrak{S}_{p,p}$ to the estimate (2).

Let us explain the use of symbols of the type $N(\cdot\,|\mathscr{E})$. $\mathscr{E}$ is always a normed or quasi-normed (q.n.) linear space and the functional $N(\cdot\,|\mathscr{E})$ constitutes a norm or quasi-norm in $\mathscr{E}$. We do not exclude the case where $\mathscr{E}$ is normable but N does not coincide with the norm and is only equivalent to it. This is in particular the case for the spaces $\mathfrak{S}_{p,q}$: they are normable for $1 < p < \infty$, $1 \leqslant q \leqslant \infty$ and for $p = q = 1$, but the functional $N(\cdot\,|\,\mathfrak{S}_{p,q})$ is a norm only for $q = p \geqslant 1$. In using q.n. classes of operators, we essentially rely on the results of Rotfel'd [3].

Let $X \subseteq \mathbf{R}^{m_1}$ and $Y \subseteq \mathbf{R}^{m_2}$ be measurable sets, and put $\mathfrak{H}_1 = L_2 X$ and $\mathfrak{H}_2 = L_2 Y$. Let the "weights" a and b and the kernel T be measurable functions on X, Y and $X \times Y$ respectively. We consider integral operators of the form

$$(\mathbf{T}_{ab}u)(y) = \int_X a(x)T(x,y)b(y)u(x)\,dx, \qquad \mathbf{T}_{ab}: \mathfrak{H}_1 \to \mathfrak{H}_2. \tag{4}$$

It is convenient to single out the weights a and b, because then we can separate the smoothness conditions (imposed on $\mathbf{T}$) and the summability conditions (imposed on a and b). In a number of cases it is sufficient to distinguish only one of the weights, say a. The operator (4) with $b \equiv 1$ will be denoted by $\mathbf{T}_a$:

$$(\mathbf{T}_a u)(y) = \int_X a(x)T(x,y)u(x)\,dx. \tag{5}$$

We first give estimates for the s-numbers of operators (4), (5) of general type, and then consider some special classes of kernels for which the results can be considerably improved.

Let us describe the spaces of functions (weights, kernels) that appear in the formulations.

1) *The Lorentz classes* $L_{p,q}X$ $(1 \leqslant p < \infty,\ 0 < q \leqslant \infty)$. Let f be a measurable a.e. finite function on X and denote by $f^*(t)$, $t > 0$, the nonincreasing function which is equimeasurable with $|f(x)|$. The relation $f \in L_{p,q}X$ means that the functional $N(f\,|\,L_{p,q}X)$ is finite, where

$$N^q(f|L_{p,q}X) = \int_0^{\operatorname{meas} X} t^{qp^{-1}-1}[f^*(t)]^q\,dt \qquad (0 < q < \infty);$$

$$N(f|L_{p,\infty}X) = \sup_t \left[t^{p^{-1}} f^*(t)\right].$$

Let us observe that $L_{p,p}X = L_pX$. For all $q \in (0, \infty]$ we have the continuous imbeddings

$$L_{p-\varepsilon}X \cap L_{p+\varepsilon}X \hookrightarrow L_{p,q}X; \qquad L_{p+\varepsilon}X \hookrightarrow L_{p,q}X \qquad (\text{meas}\, X < \infty).$$

2) *The Sobolev spaces $W_p^\alpha X$, the Nikol'skiĭ spaces $H_p^\alpha X$, and the Besov spaces* $B_{p,q}^\alpha X$ ($\alpha > 0$, $1 \leqslant p < \infty$, $0 < q \leqslant \infty$). Regarding these spaces see [4]; the case $0 < q < 1$ is in [5]. The norm in these spaces will be denoted by $N(\cdot \mid W_p^\alpha X)$, etc. Each of these norms is a sum of a "higher" term, denoted by $\mathring{N}(\cdot)$, and a "lower" term, the norm in L_pX. Let us recall that $B_{p,\infty}^\alpha = H_p^\alpha$.

Along with $W_p^\alpha \mathbf{R}^m$, $H_p^\alpha \mathbf{R}^m$ and $B_{p,q}^\alpha \mathbf{R}^m$ we define for $p\alpha < m$ the larger function spaces $\mathscr{W}_p^\alpha \mathbf{R}^m$, $\mathscr{H}_p^\alpha \mathbf{R}^m$ and $\mathscr{B}_{p,q}^\alpha \mathbf{R}^m$. For instance, $f \in \mathscr{B}_{p,q}^\alpha \mathbf{R}^m$ means that $f \in B_{p,q}^\alpha$ on every compact subset of $\mathbf{R}^m$ and that the functional (norm) $N(f|\mathscr{B}_{p,q}^\alpha \mathbf{R}^m)$ is finite, where

$$N^p\left(f|\mathscr{B}_{p,q}^\alpha \mathbf{R}^m\right) = \mathring{N}^p\left(f|B_{p,q}^\alpha \mathbf{R}^m\right) + \int_{\mathbf{R}^m} \left(1 + |x|^m\right)^{-1} |f(x)|^p\, dx. \tag{6}$$

In the same way we define the norm in $\mathscr{W}_p^\alpha \mathbf{R}^m$. A more detailed treatment of the classes just introduced will be given in Appendix 1.

3) *Spaces of functions of two variables* (*spaces of kernels*). It is convenient to interpret these classes from the standpoint of abstract functions. Let $\mathscr{E}$ be a (quasi-) normed space, $N(\cdot|\mathscr{E})$ a (quasi-) norm in $\mathscr{E}$ and $Y \subset \mathbf{R}^m$ a measurable set. By $L_rY\mathscr{E}$, $1 \leqslant r \leqslant \infty$, we denote the space of measurable maps $f\colon Y \to \mathscr{E}$ with $N(f(\cdot)|\mathscr{E}) \in L_rY$. The functional

$$N(f|L_rY\mathscr{E}) = \left[\int_Y N^r(f(y)|\mathscr{E})\, dy\right]^{1/r}, \qquad 1 \leqslant r < \infty$$

(with the usual modification if $r = \infty$) gives a quasi-norm in $L_rY\mathscr{E}$. As a rule we shall use only spaces of the type $L_rYB_{p,q}^\alpha X$.

The object of this investigation is to obtain estimates for operators $\mathbf{T}_{ab}$ and $\mathbf{T}_a$ of the forms (4) and (5) that are sharp both with respect to the rate of decrease of the s-numbers and with respect to the assumptions of smoothness and summability imposed on the kernel and the weight. In order to check sharpness we need extraneous "control sources". Let us mention the most important of the latter.

1) *Fourier series.* Let $X = Y = Q^m \overset{\text{def}}{=} [0,1]^m$, $a = b \equiv 1$ and $T(x, y) = F(x - y)$ with F periodic. Then the s-numbers of $\mathbf{T}_{ab}$ coincide up to a rearrangement with the moduli of the Fourier coefficients of F. Therefore results on the behavior of Fourier coefficients can serve as verification of the sharpness of estimates for the s-numbers with respect to classes of kernels.

2) *Differential equations.* In domain $X \subset \mathbf{R}^m$ let us consider the first (spectral) boundary value problem for the equation

$$(-1)^l \Delta^l u(x) = \lambda^{-1} g(x) u(x), \qquad g \geqslant 0.$$

For this problem one has (see [6] and [7]) sharp estimates for the spectrum:

$$\lambda_n \leqslant Cn^{-\omega} N(g|L_rX), \qquad \omega = 2l/m. \tag{7}$$

Here $r = 1$ for $\omega > 1$, $r > 1$ for $\omega = 1$, and $r = \omega^{-1}$ for $\omega < 1$. If $\omega < 1$ the domain $X \subseteq \mathbf{R}^m$ is arbitrary; if $\omega \geqslant 1$ it must be bounded. The estimate (7) is accompanied by an asymptotic formula of the type $\lambda_n \sim c(g)n^{-\omega}$.

In the applications to integral equations this corresponds to operators of the form $\mathbf{T}_{ab}$ with $Y = X$ and $b = a = \sqrt{g}$; the kernel $T(x, y)$ has a singularity of the form $|x - y|^{2l-m}$ (modified with a logarithmic factor if $2l - m \geqslant 0$ is an even integer). Thus we have the possibility of verifying the sharpness of the estimates with respect to certain classes of kernels and weights.

3) *The asymptotics of s-numbers of integral operators*. Let $X = Y \subset \mathbf{R}^m$ and let $T(x, y) = F(x - y)$ with F positive homogeneous of order $k > -m$. In this case the s-numbers of the operator $\mathbf{T}_{ab}$ of the form (4) under broad assumptions on the weights a and b admit the asymptotics

$$s_n(\mathbf{T}_{ab}) \sim c(F)\left[\int_X |a(x)b(x)|^{1/\omega}\,dx\right]^{\omega} n^{-\omega}, \qquad \omega = 1 + \frac{k}{m}. \tag{8}$$

This method of control obviously supplements the previous one. Let us observe that the derivation of (8) is itself based on estimates for the s-numbers.

Below in order to simplify the statements we always take $X = Q^m$ or, if possible, $X = \mathbf{R}^m$. The estimates have the form

$$s_n(\mathbf{T}_a) \leqslant Cn^{-\omega}N(T|L_2Y\mathscr{E})N(a|L_\lambda X) \tag{9}$$

and are initially (before the interpolation) obtained for the classes $\mathscr{E} = W_p^\alpha Q^m$ or $\mathscr{E} = \mathscr{W}_p^\alpha \mathbf{R}^m$. Here

$$\omega = \frac{1}{2} + \frac{\alpha}{m} \quad (p \geqslant 2); \qquad \omega = 1 - \frac{1}{p} + \frac{\alpha}{m} \quad (p < 2). \tag{10}$$

The constant C depends only on α, p, m and (if $p\alpha = m$) λ, but not on Y. Such a "universality" of the estimates is of importance for the applications.

THEOREM 1. *The estimate* (9)–(10) *holds in the following two cases*:
1) $X = Q^m$, $\mathscr{E} = W_p^\alpha Q^m$, $p\alpha \geqslant m$, *with* $\lambda = 2$ *if* $p\alpha > m$, *or* $\lambda > 2$ *if* $p\alpha = m$.
2) $X = \mathbf{R}^m$, $\mathscr{E} = \mathscr{W}_p^\alpha \mathbf{R}^m$, $p\alpha < m$, $p^{-1} - \alpha m^{-1} < 1/2$, *and*

$$\lambda^{-1} = 1/2 + \alpha m^{-1} - p^{-1}. \tag{11}$$

Theorem 1 is obtained on the basis of a special method of piecewise-polynomial approximation. This method is in an essential way based on the subadditivity of the norm in $W_p^\alpha X$ as a set function. Therefore it cannot be applied directly([1]) for $\mathscr{E} = B_{p,q}^\alpha X$, $p \neq q$. At the same time it turns out that, although Theorem 1 gives the best rate of decrease of the s-numbers in the scale of Sobolev spaces, the estimate (9)–(10) is not free from deficiencies. For instance, a kernel of Green's function type does not admit an exact description within the scale W_p^α, and an application of Theorem 1 to it gives for ω a number differing "by an ε" from the correct one. Next, a comparison with Fourier series leads to the hypothesis that at least for $\mathscr{E} = W_2^\alpha Q^m$ it should be $s_n = o(n^{-\omega})$.

([1])In [1] it is erroneously asserted that this method can be carried over to the spaces H_p^α.

Theorem 1 can in fact be improved in the directions indicated. It is important that for this no new analytical considerations are required: the sharpening is gotten from Theorem 1 on the basis of the general technique of interpolating linear and bilinear mappings in normed and quasi-normed spaces. Besides interpolation theorems of general character we need some new facts on bilinear interpolation specially adapted to the cases of interest to us.

§2. Interpolation methods

The sets $X \subset \mathbf{R}^{m_1}$ and $Y \subset \mathbf{R}^{m_2}$ will always be considered as fixed. Let us consider the following maps acting from spaces of functions into spaces of operators:

$$\Gamma_1 = \Gamma_1(a)\colon T \to \mathbf{T}_a; \tag{12}$$

$$\Gamma_2\colon (a, T) \to \mathbf{T}_a. \tag{13}$$

In the sequel we will likewise encounter the map

$$\Gamma_3 = \Gamma_3(T)\colon (a, b) \to \mathbf{T}_{ab}. \tag{14}$$

The map Γ_1 is linear, while Γ_2 and Γ_3 are bilinear. Below we use the notation $\|\Gamma\|$ in the following sense. If $\Gamma\colon \mathscr{E}_1 \to \mathscr{E}_2$, $\mathscr{E}_1$ and $\mathscr{E}_2$ being q.n. spaces, then

$$\|\Gamma\| = \|\Gamma\|_{\mathscr{E}_1 \to \mathscr{E}_2} \stackrel{\text{def}}{=} \sup\{N(\Gamma u|\mathscr{E}_2)\colon N(u|\mathscr{E}_1) \leqslant 1\}.$$

If $\Gamma\colon \mathscr{E}_1 \times \mathscr{E}_2 \to \mathscr{E}_3$, then

$$\|\Gamma\| = \|\Gamma\|_{\mathscr{E}_1 \times \mathscr{E}_2 \to \mathscr{E}_3} \stackrel{\text{def}}{=} \sup\{N(\Gamma(u, v)|\mathscr{E}_3)\colon N(u|\mathscr{E}_1) \leqslant 1,\ N(v|\mathscr{E}_2) \leqslant 1\}.$$

As in the normed case the finiteness of $\|\Gamma\|$ is equivalent to the continuity of the mapping Γ.

Theorem 1 can be interpreted in the following manner in terms of the maps (12) and (13). Under the hypotheses of case 1) of Theorem 1:

$$\Gamma_1(a)\colon L_2YW_p^\alpha Q^m \to \mathfrak{S}_{\delta,\infty}, \qquad \delta = \omega^{-1};$$

$$\|\Gamma_1(a)\| \leqslant C(m, \alpha, p)N(a|L_\lambda Q^m). \tag{15}$$

Under the hypotheses of case 2) of Theorem 1:

$$\Gamma_2\colon L_\lambda \mathbf{R}^m \times L_2\mathscr{W}_p^\alpha \mathbf{R}^m \to \mathfrak{S}_{\delta,\infty}, \qquad \delta = \omega^{-1}, \tag{16}$$

with $\|\Gamma_2\|$ depending on m, α, and p only.

We have observed above that further results on s-numbers are derived from here with the aid of interpolation. We employ the "real" interpolation scheme of Lions and Peetre, including the case of q.n. spaces [8], [9]. According to this scheme, to each pair of q.n. spaces (A_0, A_1) there corresponds a two-parameter family of q.n. spaces $A_{\theta,q}$, $0 < \theta < 1$, $0 < q \leqslant \infty$. If A_0 and A_1 are normed spaces and $q \geqslant 1$, then $A_{\theta,q}$ is also a normed space.

Let us now state the basic interpolation theorems. We use notation such as $\|\Gamma\|_0$, $\|\Gamma\|_1$, $\|\Gamma\|_{\theta,q}$, $\|\Gamma\|_{\theta,q_1,q_2,q}$, the meaning of which is clear from the context.

THEOREM 2 (on interpolation of linear maps [8], [10]). *Let* (A_0, A_1) *and* (B_0, B_1) *be interpolation couples of q.n. spaces and let* $\Gamma\colon A_i \to B_i$, $i = 0, 1$. *Then* $\Gamma\colon A_{\theta,q} \to B_{\theta,q}$ *for any* $\theta \in (0,1)$ *and* $q \in (0, \infty]$, *and*

$$\|\Gamma\|_{\theta q} \leqslant \|\Gamma\|_0^{1-\theta} \|\Gamma\|_1^{\theta}.$$

This theorem has the same formulation for normed as for q.n. spaces. For bilinear maps the state of affairs is different, and we must introduce a new concept. Assume that for some r, $0 < r \leqslant 1$, the quasi-norm in a q.n. space $\mathscr{E}$ satisfies the inequality

$$N^r(u + v|\mathscr{E}) \leqslant N^r(u|\mathscr{E}) + N^r(v|\mathscr{E}).$$

We then say that the space $\mathscr{E}$ is *r-normed* and that the functional N is an *r-norm*. Let us remark that in every q.n. space there exist r-norms equivalent to the basic quasi-norm.*

THEOREM 3 [11]. *Let* (A_0, A_1), (B_0, B_1) *and* (C_0, C_1) *be interpolation couples of q.n. spaces, assuming that* C_i *is* r_i*-normed*, $i = 0, 1$. *Let there be given a continuous bilinear map* $\Gamma\colon A_i \times B_i \to C_i$, $i = 0, 1$. *Let* $0 < \theta < 1$, $r_\theta^{-1} = (1-\theta) r_0^{-1} + \theta r_1^{-1}$ *and* $q_1^{-1} + q_2^{-1} \geqslant r_\theta^{-1}$. *Then* $\Gamma\colon A_{\theta, q_1} \times B_{\theta, q_2} \to C_{\theta, q}$, *where*

$$\left.\begin{aligned} q &= \max(q_1, q_2), && \text{if } \min(q_1, q_2) < r_\theta, \\ q^{-1} &= q_1^{-1} + q_2^{-1} - r_\theta^{-1}, && \text{if } \min(q_1, q_2) \geqslant r_\theta. \end{aligned}\right\} \tag{17}$$

Also,

$$\|\Gamma\|_{\theta, q_1, q_2, q} \leqslant c \|\Gamma\|_0^{1-\theta} \|\Gamma\|_1^{\theta}. \tag{18}$$

The case where the spaces A, B and C are normed (in particular, then $r_0 = r_1 = r_\theta = 1$) and q_1, $q_2 \geqslant 1$ corresponds to the Lions-Peetre theorem [8].

Formula (17) contains bounds which become stronger if we decrease r_0 and r_1. In this connection it is of importance to know the maximal value $r = r_{\max}$ for each given q.n. space. For the classes $\mathfrak{S}_{p,q}$ with $\min(p, q) < 1$ it follows from the results of Rotfel'd [3] that $r_{\max} = \min(p, q)$. The space $\mathfrak{S}_{1,q}$ for $1 < q \leqslant \infty$ admits an r-norm for any $r < 1$ (but not for $r = 1$).

In the applications to the estimates of s-numbers Theorem 3 is insufficient, because there we have not taken into account certain features of the maps (13) and (14).

Consider a bilinear map $\Gamma\colon A \times B \to C$, where A is a space of measurable functions and C a space of operators. Assume that for any a_1, $a_2 \in A$ with disjoint supports and any b_1, $b_2 \in B$ we have

$$[\Gamma(a_1, b_1)]^* \Gamma(a_2, b_2) = 0. \tag{19}$$

**Translator's note*. This is the Aoki–Rolewicz theorem; r depends on the space, of course.

Then Γ will be called *semi-orthogonal*. The same word will be used if instead of (19) we have

$$\Gamma(a_1, b_1)[\Gamma(a_2, b_2)]^* = 0. \tag{19*}$$

If B too is a space of measurable functions and for $a_1 a_2 = 0$ and $b_1 b_2 = 0$ we have at the same time both equalities (19) and (19)*, then we say that Γ is *orthogonal*. It is easy to see that the mapping (13) is semi-orthogonal while (14) is orthogonal.

Below we state theorems on interpolation of semi-orthogonal and orthogonal bilinear maps. Let us agree to say that (A_0, A_1) is an admissible couple of Lorentz spaces if $A_i = L_{p_i, q_i} X$ with $p_0 \neq p_1$.

THEOREM 4. *Let (A_0, A_1) be an admissible couple of Lorentz spaces and (B_0, B_1) an arbitrary interpolation couple of q.n. spaces. Let $0 < \delta_0, \delta_1 < 2$, $\delta_0 \neq \delta_1$, and consider a continuous semi-orthogonal bilinear map Γ: $A_i \times B_i \to \mathfrak{S}_{\delta_i, \infty}$, $i = 0, 1$. Let $0 < \theta < 1$, $\delta^{-1} = (1-\theta)\delta_0^{-1} + \theta\delta_1^{-1}$, $q_1^{-1} + q_2^{-1} \geqslant \delta^{-1}$, $(q_1, q_2) \neq (\delta, \infty)$. Then Γ: $A_{\theta, q_1} \times B_{\theta, q_2} \to \mathfrak{S}_{\delta, q}$, where*

$$\left.\begin{array}{lll} q = \max(q_1, q_2), & \textit{if } \min(q_1, q_2) < \delta, \textit{ or } \delta = q_2 < q_1; \\ q > q_2, & \textit{if } \delta = q_1 \leqslant q_2; \\ q^{-1} = q_1^{-1} + q_2^{-1} - \delta^{-1}, & \textit{if } \min(q_1, q_2) > \delta. \end{array}\right\} \tag{20a}$$

Moreover, the estimate (18) *holds.*

THEOREM 5. *Let (A_0, A_1) and (B_0, B_1) be admissible couples of Lorentz spaces. Let $0 < \delta_0, \delta_1 < \infty$, $\delta_0 \neq \delta_1$, and consider a continuous orthogonal bilinear map Γ: $A_i \times B_i \to \mathfrak{S}_{\delta_i, \infty}$, $i = 0, 1$. Let $0 < \theta < 1$ and $q_1^{-1} + q_2^{-1} > \delta^{-1} = (1-\theta)\delta_0^{-1} + \theta\delta_1^{-1}$. Then Γ: $A_{\theta, q_1} \times B_{\theta, q_2} \to \mathfrak{S}_\delta$ where*

$$\left.\begin{array}{ll} q = \max(q_1, q_2), & \textit{if } \min(q_1, q_2) < \delta,\ q_1 \neq q_2; \\ q > q_1, & \textit{if } q_1 = q_2 < \delta; \\ q > \max(q_1, q_2), & \textit{if } \min(q_1, q_2) = \delta; \\ q^{-1} = q_1^{-1} + q_2^{-1} - \delta^{-1}, & \textit{if } \min(q_1, q_2) > \delta. \end{array}\right\} \tag{20b}$$

Moreover, the estimate (18) *holds.*

If $\delta_0, \delta_1 > 1$ then $\theta > 1$; consequently the restrictions on q_1 and q_2 in Theorems 4 and 5 are weaker than the ones in Theorem 3. The formulas (20) give a better value for q than (17). An application of Theorems 4 and 5 to estimates of s-numbers gives sharp (definitive) results in many cases.

Let us also state a result analogous to Theorem 4 pertaining to the case where the image spaces coincide.

THEOREM 6. *Let (A_0, A_1) be an admissible couple of Lorentz spaces and (B_0, B_1) an arbitrary interpolation couple of q.n. spaces. Let $0 < \delta < 2$ and $0 < t \leqslant \infty$, and consider a continuous semi-orthogonal bilinear map Γ: $A_i \times B_i \to \mathfrak{S}_{\delta, t}$, $i = 0, 1$. Then with $\kappa = \min(\delta, t)$ and for any $\theta \in (0, 1)$ and q_1 and q_2 such* *that*

$$q_1^{-1} + q_2^{-1} \geqslant \kappa^{-1}, \qquad (q_1, q_2) \neq (\kappa, \infty),$$

we have $\Gamma\colon A_{\theta, q_1} \times B_{\theta, q_2} \to \mathfrak{S}_{\delta, t}$ *along with an estimate of the type* (18).

An analogous result (the formulation of which we omit) also holds for orthogonal maps.

§3. Sharpened estimates for the s-numbers

The interpolation estimates (9) lead to inequalities of the type

$$N(\mathbf{T}_a | \mathfrak{S}_{\delta,\beta}) \leqslant CN(T|L_2Y\mathscr{E})N(a|L_{\lambda,\eta}X), \tag{21}$$

where $\mathscr{E} = B^{\alpha}_{p,q}Q^m$ or $\mathscr{E} = \mathscr{B}^{\alpha}_{p,q}\mathbf{R}^m$. Below we take $\delta = \omega^{-1}$ with ω given by (10).

THEOREM 7. *The estimate* (21) *holds in the following two cases*:

1) $X = Q^m$, $\mathscr{E} = B^{\alpha}_{p,q}Q^m$, $p\alpha \geqslant m$, $\beta = q \geqslant 2$, *and* $\lambda = \eta = 2$ *for* $p\alpha > m$, $\lambda = \eta > 2$ *for* $p\alpha = m$.

2) $X = \mathbf{R}^m$, $\mathscr{E} = \mathscr{B}^{\alpha}_{p,q}\mathbf{R}^m$, $p\alpha < m$, $p^{-1} - \alpha m^{-1} < 1/2$ *and* (11) *is fulfilled*; $q \geqslant 2$, $q^{-1} + \eta^{-1} \geqslant \omega$, $(q, \eta) \neq (\infty, \delta)$; $\beta = q$ *for* $\eta < \delta$, $\beta > q$ *when* $\eta = \delta$ *and* $\beta^{-1} = q^{-1} + \eta^{-1} - \omega$ *when* $\eta > \delta$.

The proof of case 1) utilizes Theorem 2, and that of case 2), Theorem 4. The basis for the interpolation is Theorem 1: we set $\alpha_0 = \alpha - \varepsilon$, $\alpha_1 = \alpha + \varepsilon$ and use (15) and (16) with these values of α and corresponding values of λ. In order to compare Theorem 7 with the initial Theorem 1, we consider two special cases of the estimate (21).

1. Let $\beta = q = \infty$. Then (21) takes the form

$$s_n(\mathbf{T}_a) \leqslant Cn^{-\omega}N(T|L_2Y\mathscr{E})N(a|L_{\lambda,\eta}X). \tag{22}$$

Here for $p\alpha > m$ we have $\mathscr{E} = H^{\alpha}_{p}Q^m$, $\lambda = \eta = 2$, for $p\alpha = m$ we have $\mathscr{E} = H^{\alpha}_{p}Q^m$, $\lambda = \eta > 2$, and for $p\alpha < m$ we have $\mathscr{E} = \mathscr{H}_p^{\alpha}\mathbf{R}^m$, (11) holds true, and $\eta < \delta$. A control based on a comparison with (7) and (8) shows that the estimate (22) is sharp with respect to the rate of decrease of the s-numbers as well as the classes of kernels. If under the hypotheses of Theorem 7 $q < \infty$, then $s_n(\mathbf{T}_a) = o(n^{-\omega})$. This follows directly from the formulation if $p\alpha \geqslant m$, and is easily verified if $p\alpha < m$. Thus the fact that one cannot improve the "individual" estimates for the s-numbers is connected with the circumstance that we are dealing with kernels belonging to Nikol′skiĭ classes.

2. Let $p = q = 2$ and $2\alpha > m$. Then (21) takes the form

$$\sum_n n^{2\alpha/m}s_n^2(\mathbf{T}_a) \leqslant C^2N^2(T|L_2YW_2^{\alpha}Q^m)N^2(a|L_2Q^m).$$

This essentially improves Theorem 1 in the case at hand. A comparison with Fourier series shows that the estimate is sharp.

Now let us consider the case of a difference kernel

$$T(x, y) = F(x - y). \tag{23}$$

Here the main interest will be directed towards a study of operators $\mathbf{T}_{ab}$ of the type (4). The reason for this is that in the kernel (23) the variables enter in a symmetric fashion and it is therefore natural to look for estimates for $s_n(\mathbf{T}_{ab})$ under symmetric conditions on the weights. If $p\alpha \geqslant m$ then it is not hard to derive from Theorem 7 results of the required nature for $X = Y = Q^m$ and $a, b \in L_\lambda Q^m$ ($\lambda = 2$ for $p\alpha > m$ and $\lambda > 2$ for $p\alpha = m$). If $p\alpha < m$ such consequences are rather crude. However, repeated interpolation makes it possible to get sharper results. In fact we have the

THEOREM 8. *Let $X = Y = \mathbf{R}^m$ and $F \in \mathscr{B}^\alpha_{p,q}\mathbf{R}^m$ for $p \geqslant 1$, $p\alpha < m, 0 < q \leqslant \infty$. Let $a \in L_{\lambda_1, r_1}\mathbf{R}^m$ and $b \in L_{\lambda_2, r_2}\mathbf{R}^m$, with*

$$2 < \lambda_1, \lambda_2 < \infty; \qquad \lambda_1^{-1} + \lambda_2^{-1} = 1 - p^{-1} + \alpha m^{-1}; \tag{24}$$

$$\left.\begin{aligned} &r_1^{-1} + r_2^{-1} \geqslant \kappa^{-1}, \quad \text{if } \kappa \overset{\text{def}}{=} \min(\delta, q) < 2, \text{ or } \delta = q = 2;\\ &r_1^{-1} + r_2^{-1} > \kappa^{-1}, \quad \text{if } \kappa \geqslant 2, (\delta, q) \neq (2,2). \end{aligned}\right\} \tag{25}$$

Then for the operator $\mathbf{T}_{ab}$ with the kernel (23) *we have*

$$N(T_{a,b}|\mathfrak{S}_{\delta,q}) \leqslant CN(F|\mathscr{B}^\alpha_{p,q}\mathbf{R}^m)N(a|L_{\lambda_1,r_1})N(b|L_{\lambda_2,r_2}). \tag{26}$$

The estimate (26) is sharp. Let us remark that if $p^{-1} - \alpha m^{-1} > 1/2$ then clearly $\delta > 2$, so that $\mathbf{T}_{ab}$ will in general not belong to the Hilbert–Schmidt class $\mathfrak{S}_2$. The hypotheses of Theorems 1 and 7 do not admit such a possibility.

Let us explain the interpolation leading to Theorem 8, limiting ourselves to the case $p^{-1} - \alpha m^{-1} < 1/2$. We apply the estimate (21) to the kernel $F(x - y)b(y)$, $b \in L_2\mathbf{R}^m$. For the operator $\mathbf{T}_{ab}$ with the kernel (23) we then get

$$N(\mathbf{T}_{ab}|\mathfrak{S}_{\delta,\beta}) \leqslant CN(F|\mathscr{B}^\alpha_{p,q}\mathbf{R}^m)N(a|L_{\lambda,\eta}\mathbf{R}^m)N(b|L_2\mathbf{R}^m). \tag{27}$$

For difference kernels this estimate can be extended from $q \geqslant 2$ to arbitrary $q > 0$. For a fixed F inequality (27) means that

$$\Gamma_3\colon L_{\lambda,\eta}\mathbf{R}^m \times L_2\mathbf{R}^m \to \mathfrak{S}_{\delta,\beta}, \qquad \|\Gamma_3\| \leqslant CN(F|\mathscr{B}^\alpha_{p,q}\mathbf{R}^m). \tag{28}$$

Here Γ_3 is the mapping (14). Interchanging the roles of the variables x and y, we get an analogous relation. Together with (28) it serves as the basis for the interpolation; the interpolation itself is founded on Theorem 6.

Theorem 8 contains the following "individual" estimates for the s-numbers:

THEOREM 8a. *Let $F \in \mathscr{H}_p^\alpha\mathbf{R}^m$, $p\alpha < m$, $a \in L_{\lambda_1}\mathbf{R}^m$ and $b \in L_{\lambda_2}\mathbf{R}^m$, and assume that* (24) *is fulfilled. Then, for the operator $\mathbf{T}_{ab}$ with kernel* (23),

$$s_n(\mathbf{T}_{ab}) \leqslant Cn^{-\omega}N(F|\mathscr{H}_p^\alpha)N(a|L_{\lambda_1})N(b|L_{\lambda_2}).$$

From Theorems 7 and 8 one can in particular get membership conditions of the type $\mathbf{T}_{ab} \in \mathfrak{S}_r$, including tests for nuclearity ($r = 1$). However, in order to get nuclearity criteria it is more efficient to use a different approach (see, for example, [12]). Here we state one result in this direction. Let $W_2^{\alpha,\beta}\mathbf{R}^m$ denote the completion of the class $C_0^\infty \mathbf{R}^m$ with respect to the metric form

$$N^2\big(u|W_2^{\alpha,\beta}\mathbf{R}^m\big) \overset{\text{def}}{=} \mathring{N}^2\big(u|W_2^{\alpha}\mathbf{R}^m\big) + \mathring{N}^2\big(u|W_2^{\beta}\mathbf{R}^m\big).$$

THEOREM 9. *Let $X = \mathbf{R}^m$, $a \in L_2\mathbf{R}^m$ and $T \in L_2 Y W_2^{\alpha,\beta}\mathbf{R}^m$ with $0 \leqslant \beta < m/2 < \alpha$. Then $\mathbf{T}_a \in \mathfrak{S}_1$ and*

$$N(T_a|\mathfrak{S}_1) \leqslant CN\big(T|L_2YW_2^{\alpha,\beta}\mathbf{R}^m\big)N\big(a|L_2\mathbf{R}^m\big).$$

§4. Some special cases

Here we consider some special classes of kernels of interest for the applications. Not all results pertaining to these cases follow in their full force from the general theorems of §3. However an application of a similar technique yields the desired estimates. An exception constitutes the result of §4.1, which requires somewhat different means.

1. Let X, $Y \subset \mathbf{R}^m$. We consider the kernel

$$T(x,y) = \varphi(x,y)|x-y|^k, \qquad k > -m. \tag{29}$$

The corresponding estimates for the s-numbers take the form

$$s_n(\mathbf{T}_{ab}) \leqslant Cn^{-(1+k/m)}N(\varphi|L_\infty Y_{\mathscr{E}})N\big(a|L_{\lambda_1}X\big)N\big(b|L_{\lambda_2}Y\big). \tag{30}$$

THEOREM 10. *Let $X = Q^m$ if $k \geqslant 0$ and $X = \mathbf{R}^m$ if $-m < k < 0$; let $\mathscr{E} = H_p^s X$, $p \geqslant 2$, $ps > m$, $s > k + m/2$ if $k \geqslant -m/2$, and $\mathscr{E} = L_\infty X$ if $-m < k < -m/2$. Moreover assume that the following conditions hold*:

1) $\lambda_1 = \lambda_2 = 2$ *for* $k > 0$,
2) $\lambda_1 > 2$, $\lambda_2 = 2$ *or* $\lambda_1 = 2$, $\lambda_2 > 2$ *for* $k = 0$,
3) $2 < \lambda_1, \lambda_2 < \infty$ *and* $\lambda_1^{-1} + \lambda_2^{-1} = 1 + km^{-1}$ *for* $k < 0$.

Then the estimate (30) *holds for the operator* $\mathbf{T}_{ab}$ *with kernel* (29).

For $k > -m/2$ the proof is based on the fact that for $\alpha = k + m/2$ the function $|z|^k$ belongs to the class $H_2^\alpha Q^m$ ($k \geqslant 0$) or to the class $\mathscr{H}_2^\alpha \mathbf{R}^m$ ($-m/2 < k < 0$). Therefore the estimate (30) with $\varphi \equiv 1$ follows directly from Theorems 7 and 8a. The conditions on φ in Theorem 10 are such that the inclusion of φ into the kernel does not worsen the order of the estimates (regarding this approach see §5, where the problem of multipliers in kernel classes is discussed). We emphasize that for $k < -m/2$ no smoothness conditions at all are needed on φ. This important fact was in essence established by Kostometov [13].

The estimate (30) can be extended to the case of the kernel $\varphi(x,y)F(x-y)$, where F is any positive homogeneous function of degree k. The same estimate holds also for kernels of the type $\varphi(x,y)P(x-y)\log|x-y|$, where P is a polynomial of degree $k \geqslant 0$.

A comparison with (7) and (8) shows that these results are sharp with respect to the order of the estimates as well as to the weight classes.

2. If the kernel T has singularities only on the boundary, then compared to the estimate (30) the s-numbers decrease faster. Let us illustrate this by the example of an operator in $L_2 Q^m$ with the kernel

$$B(x, y) = \left[|x' - y'|^2 + |x'' + y''|^2\right]^{k/2}, \qquad k > -m. \tag{31}$$

Here $x', y' \in Q^{m_1}$ and $x'', y'' \in Q^{m_2}$, $m_1 + m_2 = m$, $m_1 > 0$, $m_2 \geqslant 0$. For the s-numbers of an operator $\mathbf{B}$ with kernel (31) we can obtain the estimate $s_n(\mathbf{B}) = o(n^{-(m+k)m_1^{-1}})$ for $a = b \equiv 1$. For the kernel $B(x, y) = |x + y|^{-k}$ (that is, the case $m_1 = 0$) the estimate takes the form $s_n(\mathbf{B}) = o(e^{-\eta\sqrt{n}})$, $\eta = \eta(m, k) > 0$.

3. In the space $L^2\mathbf{R}^m$ let us consider the operator $\mathbf{E}_{gh}$ with kernel

$$E_{gh}(x, y) = g(x)h(y)\exp(ixy).$$

One can give convenient conditions which assure that it belongs to classes $\mathfrak{S}_r$, $0 < r < \infty$. Let us define the function spaces G_r, $0 < r < 2$, which appear in the formulation. Let $\mathbf{R}^m$ be decomposed into a network of unit cubes $\{Q_k\}$, $k = 1, 2, \ldots$. Then $f \in G_r$ means that $f \in L_{2,\mathrm{loc}}\mathbf{R}^m$ and

$$N^r(f|G_r) \overset{\text{def}}{=} \sum_k N^r(f|L_2 Q_k) < \infty.$$

THEOREM 11. *Let g, $h \in \mathscr{E}_r$, where $\mathscr{E}_r = L_2\mathbf{R}^m$ if $2 \leqslant r < \infty$ and $\mathscr{E}_r = G_r$ if $0 < r < 2$. Then $\mathbf{E}_{gh} \in \mathfrak{S}_r$ and*

$$N(\mathbf{E}_{gh}|\mathfrak{S}_r) \leqslant C_r N(g|\mathscr{E}_r)N(h|\mathscr{E}_r).$$

4. Consider the operator $\mathbf{A}$ in $L_2\mathbf{R}^m$ with kernel

$$A(x, y) = \left(1 + |x|^2\right)^{-\alpha/2}\left(1 + |y|^2\right)^{-\beta/2}\left(1 + |x - y|^2\right)^{-\gamma/2} \qquad (\gamma > 0). \tag{32}$$

It is easy to see that the conditions $\alpha + \beta > 0$, $\gamma + \alpha + \beta > m$, $2(\gamma + \alpha) > m$ and $2(\gamma + \beta) > m$ are necessary for the compactness of $\mathbf{A}$. Here one of the numbers α or β may be negative; that is, the kernel may contain an increasing factor. Under these assumptions on the exponents we have the estimate

$$s_n(\mathbf{A}) = o(n^{-(\alpha+\beta)/m}). \tag{33}$$

This estimate will be proved in Appendix 2.

§5. Multipliers in classes of integral kernels

The material of this section allows us to extend considerably the range of applicability of the estimates of §§3 and 4. On the other hand it is of use also in the solution of a number of questions not directly related to estimates of s-numbers. However, we will not discuss the latter here, instead referring the reader to our lectures [2]. There the very statement of the problem of multipliers is also set forth in greater detail.

1. Let $X \subset \mathbf{R}^{m_1}$ and $Y \subset \mathbf{R}^{m_2}$, and consider a measurable function φ on $X \times Y$. We associate with it a linear map M_φ over operators acting from $L_2 X$ into $L_2 Y$. Namely if $\mathbf{T}$ is an integral operator with kernel T then $\mathbf{Q} = M_\varphi \mathbf{T}$ will be the integral operator with kernel φT:

$$(\mathbf{T}u)(y) = \int_X T(x, y)u(x)\,dx;$$

$$(\mathbf{Q}u)(y) = \int_X \varphi(x, y)T(x, y)u(x)\,dx.$$

The transformation M_φ will be called the *multiplier transformation*. Let us fix any of the classes $\mathfrak{R}$, $\mathfrak{S}_\infty$ or $\mathfrak{S}_{p,q}$, which we denote by $\mathfrak{S}$. Let $\mathcal{M}(\mathfrak{S})$ be the set of functions φ such that M_φ maps $\mathfrak{S}$ continuously into itself. If $\varphi \in \mathcal{M}(\mathfrak{S})$ then the map M_φ, defined initially on integral operators $T \in \mathfrak{S}$ only, can be extended to the whole class $\mathfrak{S}$.

The set $\mathcal{M}(\mathfrak{S})$ is an algebra, and a quasi-norm $[\varphi]\mathfrak{S}$ (or a norm, if $\mathfrak{S}$ is normed) is defined in it in a natural way. Among these the space $\mathcal{M} \overset{\text{def}}{=} \mathcal{M}(\mathfrak{R})$ is the most important for the applications. Its norm will be denoted by $[\varphi]$.

The problem of multipliers in kernel classes consists in the analytical description of the spaces $\mathcal{M}(\mathfrak{S})$. Clearly $\mathcal{M}(\mathfrak{S}_2) = L_\infty(X \times Y)$. For other classes $\mathcal{M}(\mathfrak{S})$ an explicit description was not obtained. Let us mention a few relations of general nature:

1) $\mathcal{M}(\mathfrak{S}_1) = \mathcal{M}(\mathfrak{S}_\infty) = \mathcal{M}$, and $[\varphi]\mathfrak{S}_1 = [\varphi]\mathfrak{S}_\infty = [\varphi]$.

2) If $0 < t, p, q \leqslant \infty$ and $|p^{-1} - 1/2| < |t^{-1} - 1/2|$, then the following continuous imbeddings hold:

$$\mathcal{M}(\mathfrak{S}_t) \hookrightarrow \mathcal{M}(\mathfrak{S}_{p,q}) \hookrightarrow \mathcal{M}(\mathfrak{S}_2) = L_\infty(X \times Y). \tag{34}$$

In particular, for $1 < p < \infty$ and $0 < q \leqslant \infty$

$$\mathcal{M} \hookrightarrow \mathcal{M}(\mathfrak{S}_{p,q}) \hookrightarrow L_\infty(X \times Y).$$

The analytical results on multipliers are based on the following theorem. Let $\boldsymbol{\varphi}_{ab}$ denote the operator of type (4) with kernel φ.

THEOREM 12. *If* $0 < r \leqslant 1$, *then*

$$[\varphi]\mathfrak{S}_r = \sup_{N(a|L_2X) = N(b|L_2Y) = 1} N(\boldsymbol{\varphi}_{ab}|\mathfrak{S}_r).$$

In particular, $[\varphi] = \sup N(\boldsymbol{\varphi}_{ab}|\mathfrak{S}_1)$.

Theorem 12 makes it possible to apply the results of §3 in the study of multipliers. This explains also why it is so important to have estimates uniform in the "weights" a and b.

THEOREM 13. 1) *Let* $X = \mathbf{R}^m$ *and* $0 \leqslant 2\beta < m < 2\alpha$. *Then*

$$L_\infty Y W_2^{\alpha,\beta}\mathbf{R}^m \hookrightarrow \mathcal{M}.$$

2) *Let* $X = Q^m$, $2\alpha > m$ *and* $r^{-1} < \alpha m^{-1} + 1/2$. *Then*

$$L_\infty Y H_2^\alpha Q^m \hookrightarrow \mathcal{M}(\mathfrak{S}_r).$$

3) *Let* $X = \mathbf{R}^m$, $p\alpha > m$, $|r^{-1} - 1/2| \leqslant p^{-1}$, $p \geqslant 2$. *Then*

$$L_\infty Y W_p^\alpha \mathbf{R}^m \hookrightarrow \mathcal{M}(\mathfrak{S}_r).$$

Note that under the assumptions of §3.3 certainly $r \geqslant 1$, while under the assumptions of §3.4 $r < 1$ is allowed.

Results for other classes can be obtained on the basis of the imbeddings (34). Conditions for relations of the form $\varphi \in \mathcal{M}(\mathfrak{S}_{\delta,\infty})$ are used, in particular, in the proof of Theorem 9.

Appendix 1. Interpolation relations in the classes $\mathcal{W}_p^{\alpha}$ and $\mathcal{B}_{p,q}^{\alpha}$

1. The interpolation properties of the classes W_p^{α} and $B_{p,q}^{\alpha}$ are well known, and their application does not require any further comment. In this Appendix, which has mainly a methodological character, we show that analogous interpolation properties are shared by the functional spaces $\mathcal{W}_p^{\alpha}\mathbf{R}^m$ and $\mathcal{B}_{p,q}^{\alpha}\mathbf{R}^m$ introduced in §1. We assume that $p\alpha < m$, because only this case is of any interest in the applications. On the other hand, precisely for $p\alpha < m$ there arise interesting problems in the theory of the spaces $\mathcal{W}_p^{\alpha}\mathbf{R}^m$ and $\mathcal{B}_{p,q}^{\alpha}\mathbf{R}^m$ themselves.([2])

We shall show that for the real interpolation method of Lions and Peetre we have

$$\left(\mathcal{W}_p^{\alpha_0}, \mathcal{W}_p^{\alpha_1}\right)_{\theta,q} = \mathcal{B}_{p,q}^{\alpha_0},$$

$$0 \leqslant \alpha_0 < \alpha_1 < mp^{-1}; \quad 0 < \theta < 1; \quad \alpha_\theta = (1-\theta)\alpha_0 + \theta\alpha_1;$$

$$1 \leqslant p < \infty, \qquad 0 < q \leqslant \infty. \tag{A.1}$$

The question of interpolation of the spaces $\mathcal{W}_p^{\alpha}$ and $\mathcal{B}_{p,q}^{\alpha}$ will be reduced to the analogous problem for spaces of classes of functions differing by a polynomial of fixed degree (the spaces $\mathcal{L}_p^{\alpha}$ of Sobolev and their analogues the spaces $\mathcal{L}_{p,q}^{\alpha}$). The interpolation of such spaces was considered by Peetre [14]. Here the condition $p\alpha < m$ is no longer mandatory. The reduction needed is based on the fact that for $p\alpha < m$ one can in each element (class) of the spaces $\mathcal{L}_p^{\alpha}$ and $\mathcal{L}_{p,q}^{\alpha}$ select a unique canonical representative "decreasing" as $|x| \to \infty$. Such a decrease in any case guarantees the finiteness of the integral

$$K_p(u) = \int \left(1 + |x|^m\right)^{-1} |u|^p \, dx, \tag{A.2}$$

the "lower" term of the norm (6). Similar considerations may perhaps to a considerable extent be based on the results of Sobolev [15] and Besov [16] on the density of compactly supported functions in $\mathcal{L}_p^{\alpha}$ and $\mathcal{L}_{p,q}^{\alpha}$, $q < \infty$. We prefer, however, a different (more direct) path, where in order to single out a canonical representative we will use a special projection operator onto the corresponding space of polynomials. In this way the case of the space $\mathcal{H}_p^{\alpha} = \mathcal{B}_{p,\infty}^{\alpha}$ does not require a special treatment.

([2])In what follows the symbol $\mathbf{R}^m$ will often be left out in the notation for the spaces or the (quasi-) norms. An integral without indication of the domain of integration will be assumed to be taken over $\mathbf{R}^m$. The letter C serves to denote various constants.

2. For the space $\mathscr{W}_p^\alpha$ the higher term of the norm of type (6) $\mathring{N}(\cdot\,|\,W_p^\alpha)$ is defined for integer α by the first, and for noninteger α by the second, of the following expressions:

$$\left[\int|\nabla_\alpha u|^p\,dx\right]^{1/p}; \qquad \left[\iint\frac{|\nabla_{[\alpha]}u(x)-\nabla_{[\alpha]}u(y)|^p}{|x-y|^{m+p(\alpha-[\alpha])}}\,dx\,dy\right]^{1/p}.$$

A function u defined in $\mathbf{R}^m$ belongs to $\mathscr{W}_p^\alpha$, $p\alpha<m$, if

$$N(u|\mathscr{W}_p^\alpha)\overset{\text{def}}{=}\mathring{N}(u|W_p^\alpha)+K_p^{1/p}(u)<\infty. \tag{A.3}$$

It is likewise convenient to include the space $\mathscr{W}_p^0$ in the scale $\mathscr{W}_p^\alpha$. Then one has to take $\mathring{N}(\cdot\,|\,W_p^0)=N(\cdot\,|L_p)$ in (A.3). The norm (A.3) is for $\alpha=0$ equivalent to the standard norm in L_p.

Further we associate with every function $u\in L_{p,\text{loc}}\mathbf{R}^m$ its "rth order modulus of smoothness"

$$\omega_p^{(r)}(u;t)=\sup_{|h|\leqslant t}\left\{\int\left|\sum_{0\leqslant j\leqslant r}(-1)^j\binom{r}{j}u(x+jh)\right|^p dx\right\}^{1/p}, \qquad t>0.$$

Next let

$$\mathring{N}_{l_1,l_2}(u|B_{p,q}^\alpha)=\left\{\int_0^\infty\left[\frac{\omega_p^{(l_2)}(\nabla_{l_1}u;t)}{t^{\alpha-l_1}}\right]^q\frac{dt}{t}\right\}^{1/q}, \qquad 0<q<\infty;$$

$$\mathring{N}_{l_1,l_2}(u|B_{p,\infty}^\alpha)=\sup_{t>0}\frac{\omega_p^{(l_2)}(\nabla_{l_1}u;t)}{t^{\alpha-l_1}}.$$

Here $l_1+l_2>\alpha$ and $l_2\geqslant 2$ for α an integer, and $l_2\geqslant 1$ for α nonintegral. The function u belongs to the space $\mathscr{B}_{p,q}^\alpha$, $p\alpha<m$, if

$$N_{l_1,l_2}(u|\mathscr{B}_{p,q}^\alpha)\overset{\text{def}}{=}\mathring{N}_{l_1,l_2}(u|B_{p,q}^\alpha)+K_p^{1,p}(u)<\infty. \tag{A.4}$$

In the notation for spaces we do not indicate the dependence of $\mathscr{B}_{p,q}^\alpha$ on l_1 and l_2, because eventually it becomes clear that a change of l_1 or l_2 only leads to an equivalent renorming of the space. We can therefore limit ourselves to, for instance, the values $l_1=0$ and $l_2=[\alpha]+1$.

3. Now let us pass to a description of the spaces $\mathscr{L}_p^\alpha$ and $\mathscr{L}_{p,q}^\alpha$. Below we will denote by $\mathscr{P}_s$, $s>0$, the vector space of polynomials of degree $<s$ in $\mathbf{R}^m$. Consider a function u defined in $\mathbf{R}^m$, $u\in W_{p,\text{loc}}^\alpha$ ($u\in B_{p,q,\text{loc}}^\alpha$), and assume that

$$\mathring{N}(u|W_p^\alpha\mathbf{R}^m)<\infty \qquad \left(\mathring{N}_{l_1,l_2}(u|B_{p,q}^\alpha\mathbf{R}^m)<\infty\right).$$

Functions differing by a polynomial of corresponding degree r will not be distinguished. Identification (factorization) leads to the spaces $\mathscr{L}_p^\alpha$ and $\mathscr{L}_{p,q}^\alpha$, the elements of which are classes $\hat{u}=u+\mathscr{P}_r$. Here $r=\alpha$ for $\mathscr{L}_p^\alpha$, and $r=l_1+l_2$ for $\mathscr{L}_{p,q}^\alpha$; that is, it depends on the choice of the functional $\mathring{N}_{l_1,l_2}$. In $\mathscr{L}_p^\alpha$ we introduce the norm

$$N(\hat{u}|\mathscr{L}_p^\alpha)\overset{\text{def}}{=}\mathring{N}(u|W_p^\alpha), \qquad u\in\hat{u}.$$

In $\mathscr{L}^{\alpha}_{p,q}$ we define the quasi-norm (a norm if $q \geqslant 1$)

$$N_{l_1,l_2}(\hat{u}|\mathscr{L}^{\alpha}_{p,q}) \stackrel{\text{def}}{=} \mathring{N}_{l_1,l_2}(u|B^{\alpha}_{p,q}), \qquad u \in \hat{u}.$$

The spaces $\mathscr{L}^{\alpha}_{p}$ and $\mathscr{L}^{\alpha}_{p,q}$ are complete. There exists a natural isomorphism between spaces $\mathscr{L}^{\alpha}_{p,q}$ with different l_1 and l_2, and the corresponding (quasi-) norms are then equivalent. Moreover for α nonintegral the spaces $\mathscr{L}^{\alpha}_{p}$ and $\mathscr{L}^{\alpha}_{p,p}$ can be identified. We shall not indicate the inessential dependence on l_1 and l_2 in the notation for $\mathscr{L}^{\alpha}_{p,q}$. One has the interpolation relation

$$\left(\mathscr{L}^{\alpha_0}_{p}, \mathscr{L}^{\alpha_1}_{p}\right)_{\theta,q} = \mathscr{L}^{\alpha_\theta}_{p,q}$$

$$(0 \leqslant \alpha_0 < \alpha_1 < \infty;\quad 0 < \theta < 1;\quad \alpha_\theta = (1-\theta)\alpha_0 + \theta\alpha_1;\quad 1 \leqslant p < \infty,\ 0 < q \leqslant \infty), \tag{A.5}$$

which should be interpreted in the following way. The elements of all the spaces entering into (A.5) should be taken as classes of functions defined up to one and the same space of polynomials $\mathscr{P}_s$. If necessary, one should then increase "forcibly" the corresponding values of r, retaining the natural isometric isomorphism with the initial spaces. In particular such an increase of r is always necessary if $\alpha_0 = 0$ ($\mathscr{L}^0_p \stackrel{\text{def}}{=} L_p$).

Concerning the properties of the spaces $\mathscr{L}^{\alpha}_{p}$ and $\mathscr{L}^{\alpha}_{p,q}$ discussed in this subsection, see Peetre [14], [17]. As already noted, the condition $p\alpha < m$ is not mandatory here.

4. For $p\alpha < m$ we indicate a natural procedure which helps us to distinguish functions belonging to one and the same class $\hat{u}$ ($\hat{u} \in L^{\alpha}_{p}$ or $\hat{u} \in L^{\alpha}_{p,q}$). Let r be the number defined in subsection 3. In the set $\mathscr{P}_r$ we introduce a norm: to fix the ideas we set $n(f|\mathscr{P}_r) = \max_{|x|\leqslant 1}|f(x)|$. By $\mathscr{D}_\rho$ we denote the ball $|x| < \rho$ and by P_ρ the orthogonal projection in $L_2\mathscr{D}_\rho$ onto P_r. In $B^{\alpha}_{p,q}\mathscr{D}_1$ one can introduce an equivalent norm with the aid of the lower term $N(P_1 v|L_p\mathscr{D}_1)$. Then

$$\int_{D_1} |v|^p\,dx \leqslant C\left[\mathring{N}(v|B^{\alpha}_{p,q}\mathscr{D}_1) + N(P_1 v|L_p\mathscr{D}_1)\right]^p.$$

Let us apply this inequality to the function $v - P_1 v$ and perform a change of coordinates $x \mapsto \rho^{-1}x$. Then, taking account of the homogeneity of the functional $\mathring{N}(\cdot\,|B^{\alpha}_{p,q})$, we get

$$\int_{\mathscr{D}_\rho} |v - P_\rho v|^p\,dx \leqslant C\rho^{p\alpha}\mathring{N}^p(v|B^{\alpha}_{p,q}\mathscr{D}_\rho). \tag{A.6}$$

Next let $\hat{u} \in \mathscr{L}^{\alpha}_{p,q}$. For any "representative" $u \in \hat{u}$ we get from (A.6)

$$\begin{aligned}\int_{\mathscr{D}_\rho} |P_\rho u - P_{2\rho} u|^p\,dx &\leqslant C\rho^{p\alpha}\mathring{N}^p(u|B^{\alpha}_{p,q}\mathbf{R}^m)\\ &= C\rho^{p\alpha}\mathring{N}^p(\hat{u}|\mathscr{L}^{\alpha}_{p,q}).\end{aligned} \tag{A.7}$$

Setting $f_\rho = P_\rho u - P_{2\rho}u$, $\rho \geq 1$, we have by virtue of (A.7)

$$n^p(f_p|\mathscr{P}_r) \leq \max_{x \in \mathscr{D}_\rho} |f_\rho(x)|^p \leq C\rho^{-m}\int_{\mathscr{D}_p} |f_\rho(x)|^p\,dx$$

$$\leq C\rho^{p\alpha - m}\mathring{N}^p(\hat{u}|\mathscr{L}^\alpha_{p,q}). \tag{A.8}$$

Choosing $\rho = 2^n$ we see from (A.8) since $p\alpha < m$ that as $n \to \infty$ the polynomials $P_{2^n}u$ converge uniformly in $\mathscr{D}_1$ (and consequently on any compact set in $\mathbf{R}^m$) to a polynomial which we denote by $P_\infty u$. The same argument works if we replace $\hat{u} \in \mathscr{L}^\alpha_{p,q}$ by $\hat{u} \in \mathscr{L}^\alpha_p$. It is likewise easy to see that $P_\rho u \to P_\infty u$ as $\rho \to \infty$ according to any law (not only for $\rho = 2^n$). Let us remark that $P_\infty = P^2_\infty$; that is, P_∞ is a projection onto the space $\mathscr{P}_r$. We have proved

PROPOSITION 1. *Let $\hat{u} \in \mathscr{L}^\alpha_{p,q}$ or $\hat{u} \in \mathscr{L}^\alpha_p$, $p\alpha < m$. Let $u \in \hat{u}$ and in $L_2\mathscr{D}_\rho$ consider the orthogonal projection P_ρ onto $\mathscr{P}_r$. Then the polynomials $P_\rho u$ converge uniformly on every compact set to a polynomial $P_\infty u \in \mathscr{P}_r$ as $\rho \to \infty$. The operator P_∞ is a projection onto the subspace $\mathscr{P}_r$.*

The polynomials $P_\infty u$ allow us to distinguish functions u belonging to one and the same class $\hat{u}$. The representative $u \in \hat{u}$ such that $P_\infty u = 0$ will be termed the *canonical representative*. On $\mathscr{L}^\alpha_{p,q}$ and $\mathscr{L}^\alpha_p$ we define a map

$$J\colon \hat{u} \to u, \qquad u \in \hat{u}, \qquad P_\infty u = 0, \tag{A.9}$$

with the aid of which we can treat $\mathscr{L}^\alpha_{p,q}$ and $\mathscr{L}^\alpha_p$ (for $p\alpha < m$) as function spaces.* The operator J depends, in essence, only on r but not on the basic parameters α, p and q.

Below we will see that the condition $P_\infty u = 0$ selects a representative for which the integral (A.2) is always finite. But first we establish

PROPOSITION 2. *Let $u \in L_{p,\mathrm{loc}}$ and assume that the integral* (A.2) *is finite. Then as $\rho \to \infty$ the polynomials converge uniformly to zero in any ball $\mathscr{D} \subset \mathbf{R}^m$.*

PROOF. For any $\rho > \rho_0 > 0$ we have

$$\begin{aligned}\max_{|x| \leq \rho_0} |(P_\rho u)(x)|^p &\leq \max_{|x| \leq \rho} |(P_\rho u)|^p = \max_{|y| \leq 1} |P_1[u(\rho y)]|^p \\ &\leq C\int_{\mathscr{D}_1} |u(\rho y)|^p\,dy = C\rho^{-m}\int_{\mathscr{D}_\rho} |u(x)|^p\,dx,\end{aligned} \tag{A.10}$$

and so

$$\max_{|x| \leq \rho_0} |(P_\rho u)(x)|^p \leq CK_p(u). \tag{A.11}$$

This implies that it is sufficient to check the desired convergence on the dense set of compactly supported functions (dense in the metric of the integral K_p). But for such functions the convergence of $P_\rho u$ to zero follows immediately from (A.10).

**Translator's note*. Another treatment of the same problem (viz. the existence of a canonical representative), based on the Fourier transform, can be found in [23], Chapter 3.

5. Now we show that the map J defined by (A.9) establishes an isomorphism between $\mathscr{L}_p^\alpha$ and $\mathscr{W}_p^\alpha$.

PROPOSITION 3. *Let* $u \in \mathscr{L}_p^\alpha$, $p\alpha < m$, $u \in \hat{u}$. *Then the following conditions are equivalent*:

(a) $u = J\hat{u}$, *i.e.* $P_\infty u = 0$.

(b) *Hardy's inequality holds*:

$$\int |u|^p |x|^{-p\alpha} dx \leqslant CN^p(\hat{u}|\mathscr{L}_p^\alpha) = C\mathring{N}^p(u|W_p^\alpha \mathbf{R}^m). \tag{A.12}$$

(c) *The estimate*

$$K_p(u) \leqslant CN^p(\hat{u}|\mathscr{L}_p^\alpha) \tag{A.13}$$

holds.

PROOF. Clearly (b) implies (c). From (c) follows (a) in view of Proposition 2. It remains to show that (a) implies (b). Denote by Π a continuous extension operator Π: $W_p^\alpha \mathscr{D}_1 \to W_p^\alpha \mathbf{R}^m$ such that $(\Pi v)(x) = 0$ for $|x| > 2$. The function $\tilde{v} = \Pi v$ has compact support, and therefore Hardy's inequality([3]) holds:

$$\int |\tilde{v}|^p |x|^{-p\alpha} dx \leqslant C\mathring{N}^p(\tilde{v}|W_p^\alpha \mathbf{R}^m).$$

This leads to the relation

$$\int_{\mathscr{D}_1} |v|^p |x|^{-p\alpha} dx \leqslant C\left[\mathring{N}(v|W_p^\alpha \mathscr{D}_1) + N(P_1 v|L_p \mathscr{D}_1)\right]^p.$$

Setting $v = u - P_1 u$ and performing the substitution $x \mapsto \rho^{-1}x$, we get

$$\int_{\mathscr{D}_\rho} |u - P_\rho u|^p |x|^{-p\alpha} dx \leqslant C\mathring{N}^p(u|W_p^\alpha \mathscr{D}_\rho)$$

$$\leqslant CN^p(\hat{u}|\mathscr{L}_p^\alpha). \tag{A.14}$$

Taking account of the fact that $P_\rho u \to P_\infty u$ and $P_\infty u = 0$, we get from (A.14), using Fatou's lemma, the desired estimate (A.12).

Let us emphasize that Hardy's inequality in the form (A.12) differs from its standard variant in that the condition on the behavior of $u(x)$ as $|x| \to \infty$ is reduced to the considerably weaker requirement $P_\infty u = 0$.

A simple juxtaposition of Propositions 1 and 3 yields the following result.

PROPOSITION 4. *The operator* J, *as defined by* (A.9), *establishes an isomorphism between the spaces* $\mathscr{L}_p^\alpha$ *and* $\mathscr{W}_p^\alpha$, *and, by virtue of* (A.13), *the norms in both spaces are equivalent*.

([3])Concerning Hardy's inequality for noninteger α, see [18], [19] where the case $m = 1$ is treated. For $m > 1$ a proof due to V. A. Solonnikov can be found in [2].

6. Next let us establish the analogue of Proposition 4 for the spaces $\mathscr{L}^{\alpha}_{p,q}$ and $\mathscr{B}^{\alpha}_{p,q}$. We use the fact that the isomorphism (A.9) allows us to regard the interpolation formula (A.5) as a relation between spaces of functions (and not classes of functions). From (A.5), (A.13) and the interpolation Theorem 2 it then follows directly that

$$K_p(u) \leqslant C\mathring{N}^p(u|B^{\alpha}_{p,q}\mathbf{R}^m) = CN^p(\hat{u}|\mathscr{L}^{\alpha}_{p,q}), \qquad u = J\hat{u}. \qquad \text{(A.15)}$$

From Proposition 1 and (A.15) we get

PROPOSITION 5. *The operator J, as defined in* (A.9), *establishes an isomorphism between the spaces $\mathscr{L}^{\alpha}_{p,q}$ and $\mathscr{B}^{\alpha}_{p,q}$, $p\alpha < m$, and the metrics in these spaces are equivalent.*

From this it follows in particular that the different metrics (A.4) in $\mathscr{B}^{\alpha}_{p,q}$ are equivalent, because a similar statement is true for the spaces $\mathscr{L}^{\alpha}_{p,q}$.

Finally, from Propositions 4 and 5 and the relation (A.5) we immediately deduce the

THEOREM. *The interpolation relations* (A.1) *are true.*

In conclusion we make a few remarks. The class $C_0^{\infty}\mathbf{R}^m$ is dense in $\mathscr{W}^{\alpha}_p$ and in $\mathscr{B}^{\alpha}_{p,q}$, $q \neq \infty$. This follows at once from the results of [15] and [16]. For the spaces $\mathscr{W}^{\alpha}_p$, $p\alpha < m$, this fact can be gotten easily with the aid of Hardy's inequality in the form (A.12). It can be carried over to the spaces $\mathscr{B}^{\alpha}_{p,q}$ with the aid of (A.1).

For the spaces $\mathscr{B}^{\alpha}_{p,q}$ inequalities much stronger than (A.15) hold. Corresponding results are analogous to Hardy's inequality (A.12) and can be obtained from the latter on the basis of the results of [17] and [20] on interpolation in weighted L_p-classes. For example, if $q = \infty$ (that is, for the class $\mathscr{H}^{\alpha}_p$) we have

$$\sup \frac{\int |u(x)|^p \varphi(|x|)\,dx}{\int |x|^{p\alpha - m}\varphi(|x|)\,dx} \leqslant C\mathring{N}^p(u|H^{\alpha}_p\mathbf{R}^m) \qquad (P_{\infty}u = 0),$$

where the supremum is taken with respect to all positive nonincreasing functions φ for which the integral in the denominator is finite.

The advantage of the weakest lower term (among all possible) in (A.3) and (A.4) is that it works for the entire scale of spaces $\mathscr{W}^{\alpha}_p$ and $\mathscr{B}^{\alpha}_{p,q}$.

Appendix 2

Here we give the proof of the estimate (33) for the s-numbers of the integral operator $\mathbf{A} = \mathbf{A}_{\alpha\beta}$ with kernel (32). This example is of interest, because the result in question does not follow from the theorems set forth in the main text. Recall that the operator $\mathbf{A}$ is compact in $L_2\mathbf{R}^m$ only under the conditions

$$\alpha + \beta > 0, \quad \alpha + \beta + \gamma > m, \quad \alpha + \gamma > m/2, \quad \beta + \gamma > m/2. \qquad \text{(A.16)}$$

It will be convenient to consider the number $\gamma > 0$ as fixed; the domain in which the parameters (α, β) vary, determined by (A.16), will be denoted by S_γ. The set of points (α, β) for which (33) holds will be denoted by S^0_γ. It turns out that $S_\gamma = S^0_\gamma$.

1. It is clear that the points (α, β) and (β, α) belong to S_γ^0 at the same time. Next we establish that S_γ^0 is convex. Let $(\alpha_0, \beta_0), (\alpha_1, \beta_1) \in S_\gamma^0$. Then $(\alpha, \beta) \in S_\gamma^0$, where $\alpha = (1-\kappa)\alpha_0 + \kappa\alpha_1$ and $\beta = (1-\kappa)\beta_0 + \kappa\beta_1$, $0 < \kappa < 1$. In fact, set $\mathbf{T} = \mathbf{A}_{\alpha_0, \beta_0}$ and let J_1 and J_2 be the operators of multiplication by the functions $\varphi_{\beta_1 - \beta_0}$ and $\varphi_{\alpha_1 - \alpha_0}$ respectively, with $\varphi_\alpha(x) = (1 + |x|^2)^{-\alpha/2}$. Then $J_1 \mathbf{T} J_2 = \mathbf{A}_{\alpha_1 \beta_1}$, $J_1^\kappa \mathbf{T} J_2^\kappa = \mathbf{A}_{\alpha\beta}$, and the desired estimate (33) follows directly from the interpolation Lemma 1.5 in [21].

A subset $S \subset S_\gamma$ will be *generating* if S_γ coincides with the convex hull of the union of S and the reflection of S in the line $\alpha = \beta$. It is clear that $S_\gamma = S_\gamma^0$ as soon as we have established (33) for at least one generating set S.

Together with $\mathbf{A}_{\alpha\beta}$ we consider the unitarily equivalent operator

$$\mathbf{B}_{\alpha\beta} = (I - \Delta)^{-\beta/2} G_\gamma (I - \Delta)^{-\alpha/2}.$$

Here G_γ is the operator of multiplication by the Fourier transform g_γ of φ_γ. As $|x| \to 0$ one has (see [4])

$$g_\gamma(x) = \begin{cases} O\left(|x|^{-(m-\gamma)}\right), & \gamma < m; \\ O\left(\log |x|^{-1}\right), & \gamma = m; \\ O(1), & \gamma > m, \end{cases} \tag{A.17}$$

while g_γ decays exponentially as $|x| \to \infty$.

2. Let $\beta = 0$; this case will be one of the "supporting" cases. By virtue of (A.16), if $\beta = 0$, then necessarily $\gamma > m/2$ and $\alpha > 0$. The numbers $s_n^2(\mathbf{B}_{\alpha 0})$ coincide with the successive maxima (the spectrum) of the ratio of quadratic forms

$$N^2(g_\gamma u | L_2 \mathbf{R}^m) / N^2(u | W_2^\alpha \mathbf{R}^m). \tag{A.18}$$

Assume first $2\alpha < m$. Then, by virtue of (A.17), $g_\gamma^2 \in L_{m/2\alpha}$ and consequently it is possible to apply to the spectrum of the ratio (A.18) the estimates in [7] (see also [2], where the case of nonintegral α is clarified). These estimates at once give us (33) with $\beta = 0$.

If $2\alpha \geqslant m$ then, in view of what was said in subsection 1, we can assume that α is an integer. Under these assumptions the estimate (33) with $\beta = 0$ follows on the basis of (A.17) directly from Lemma 2.1 in [22].

3. Now let us consider the case where $2\gamma \leqslant m$ (and consequently $\alpha, \beta > 0$). It is easy to see that the subset $\tilde{S}_\gamma$ of points $(\alpha, \beta) \in S_\gamma$ such that either $\alpha, \beta < m/2$ or α and β are integers greater than $m/2$ is, in the case at hand, a generating set. It is thus sufficient to prove (33) for $(\alpha, \beta) \in \tilde{S}_\gamma$. As in subsection 2, it is convenient to replace the operator $\mathbf{A}_{\alpha\beta}$ by $\mathbf{B}_{\alpha\beta}$.

We use the inequality

$$s_{2n-1}(\mathbf{B}_{\alpha\beta}) \leqslant s_n\left(G_\gamma^\varepsilon (I - \Delta)^{-\alpha/2}\right) s_n\left(G_\gamma^{1-\varepsilon} (I - \Delta)^{-\beta/2}\right), \tag{A.19}$$

where $\varepsilon > 0$ will be chosen later. As in subsection 2, we arrive at the problem of estimating the spectrum of a ratio of quadratic forms:

$$\frac{\int g_\gamma^{2\varepsilon}|u|^2\,dx}{N^2(u|W_2^\alpha\mathbf{R}^m)}, \qquad \frac{\int g_\gamma^{2(1-\varepsilon)}|u|^2\,dx}{N^2(u|W_2^\beta\mathbf{R}^m)}. \tag{A.20}$$

Assume first that $\alpha, \beta < m/2$; we choose ε such that $m-\gamma-\beta < \varepsilon(m-\gamma) < \alpha$. Then $g_\gamma^{2\varepsilon} \in L_{m/2\alpha}\mathbf{R}^m$ and $g_\gamma^{2(1-\varepsilon)} \in L_{m/2\beta}\mathbf{R}^m$. On the basis of [2] and [7] we conclude that the eigenvalues of the ratios (A.20) have the orders $O(n^{-2\alpha/m})$ and $O(n^{-2\beta/m})$ respectively. From this and (A.19), (33) follows.

If α and β are integers greater than $m/2$, we choose ε so that $m-2\gamma < 2(m-\gamma)\varepsilon < m$. Then the functions $g_\gamma^{2\varepsilon}$ and $g_\gamma^{2(1-\varepsilon)}$ are summable and decay exponentially; as in subsection 2 (the case $2\alpha \geqslant m$), the estimate (33) is obtained if we refer to Lemma 2.1 in [22].

4. The results of subsections 1–3 allow us for any $\gamma > 0$ to exhaust the domain $\alpha, \beta \geqslant 0$. If $2\gamma > m$, values $\beta < 0$ are permitted. Let us show that then it is possible to make a reduction to the case $-1 < \beta \leqslant 0$. In fact the numbers $s_n^2(\mathbf{B}_{\alpha\beta})$ coincide with the spectrum of the ratio of quadratic forms

$$\frac{N^2(g_\gamma u|W_2^{|\beta|}\mathbf{R}^m)}{N^2(u|W_2^\alpha\mathbf{R}^m)}. \tag{A.21}$$

The conditions (A.16) now mean that

$$\alpha > |\beta|, \quad \gamma - |\beta| > m/2, \quad \alpha + \gamma - |\beta| > m. \tag{A.22}$$

The numerator of the ratio (A.21) can be estimated from above by a sum of terms of the type

$$N^2(\mathscr{D}^\sigma g_\gamma \mathscr{D}^\tau u|W_2^\mu\mathbf{R}^m); \qquad |\sigma| + |\tau| = \delta,$$

where δ is the integer part of $|\beta|$, $\mu = |\beta| - \delta$, and σ and τ are multi-indices. The function $g_{\gamma,\sigma} = \mathscr{D}^\sigma g_\gamma$ satisfies conditions of the form (A.17) with γ replaced by $\gamma - |\sigma|$. Setting $\mathscr{D}^\gamma u = v$, we can reduce the problem for each term to a study of the spectrum of the ratio

$$\frac{N^2(g_{\gamma,\sigma}v|W_2^\mu\mathbf{R}^m)}{N^2(v|W_2^{\alpha-|\tau|}\mathbf{R}^m)}. \tag{A.23}$$

We have arrived at a series of problems of the type (A.21) where $|\beta|$ is replaced by μ, α by $\alpha - |\tau|$ and γ by $\gamma - |\sigma|$. Condition (A.22) is then preserved, and the problem is reduced to the case $-1 < \beta \leqslant 0$. As soon as we have obtained the estimate (33) for these β it will then carry over to any admissible $\beta < 0$. In addition the "worst" order, which determines the rate of decrease of the numbers $s_n(\mathbf{B}_{\alpha\beta})$, corresponds to the value $\sigma = 0$.

Let us remark that if β is an integer then $\mu = 0$ and the spectrum of any of the ratios (A.23) can be estimated as in subsection 2. It is clear that this at once leads to the estimate (33) for integral $\beta < 0$.

5. Let $-1 < \beta < 0$. If moreover $\gamma - m/2 > 1$ then (33) can be obtained by interpolation between suitable points on the lines $\beta = 0$ and $\beta = -1$. It remains to consider the case $0 < \gamma - m/2 \leqslant 1$ with the value $\beta = -1$ excluded. Let us write the kernel (32) in the form

$$\varphi_{\alpha-|\beta|}(x)\varphi_\gamma(x-y) + \varphi_\alpha(x)\varphi_{\gamma-|\beta|}(x-y)\left[\frac{\varphi_\beta(y)-\varphi_\beta(x)}{\varphi_\beta(x-y)}\right]. \quad \text{(A.24)}$$

Let us assume that

$$\alpha - |\beta| \leqslant m/2, \qquad \alpha > m/2. \quad \text{(A.25)}$$

Then the second term in (A.24) is the kernel of an operator of class $\mathfrak{S}_2$ (observe that the factor within square brackets is bounded for $-1 \leqslant \beta < 0$) and therefore its s-numbers admit the estimate $s_n = o(n^{-1/2})$. The first term in (A.24) is the kernel of an operator considered in subsection 2, and its s-numbers satisfy (33). In view of the first of the conditions (A.25) this determines the rate of decrease of the s-numbers of the operator $\mathbf{A}_{\alpha\beta}$ with kernel (A.24). Thus (33) is established under the auxiliary hypotheses (A.25). It remains to remark that in the case under consideration the conditions (A.25) distinguish a set in S_γ which together with the ray $\beta = 0$, $\alpha > m/2$ generates all of S_γ.

In conclusion let us remark that for $\gamma > m$ one can give a much shorter proof of (33). This is based on the observation that for $\alpha + \beta = 0$ and $\alpha, \beta > m/2 - \gamma$ the operator $\mathbf{A}_{\alpha\beta}$ is bounded and that this segment together with the ray $\beta = 0$, $\alpha > 0$ generates S_γ. Unfortunately this approach does not extend to the case $\gamma \in (m/2, m]$.

REFERENCES

1. M. Sh. Birman and M. Z. Solomyak, *Piecewise-polynomial approximations of functions of the classes* W_p^α, Mat. Sb. **73 (115)** (1967), 31–355; English transl. in Math. USSR Sb. **2** (1967).

2. ______, *Qualitative analysis in Sobolev imbedding theorems and applications to spectral theory*, Tenth Math. Summer School (Katsiveli/Nalchik, 1972), Izdanie Inst. Mat. Akad. Nauk Ukrain. SSR, Kiev, 1974, pp. 5–189; English transl., Amer. Math. Soc. Transl. (2) **114** (1980).

3. S. Yu. Rotfel'd, *The singular numbers of a sum of completely continuous operators*, Problemy Mat. Fiz., vyp. 3, Izdat. Leningrad. Univ., Leningrad, 1968, pp. 81–87; English transl. in Topics in Math. Phys., no. 3, Plenum Press, New York, 1969.

4. S. M. Nikol'skiĭ, *Approximation of functions of several variables and imbedding theorems*, "Nauka", Moscow, 1969; English transl., Springer-Verlag, 1974.

5. G. E. Karadzhov, *Extensions of functions of the Besov spaces* $B_N^\alpha L^p$, C.R. Acad. Bulgare Sci. **27** (1974), 603–606. (Russian)

6. M. Sh. Birman and M. Z. Solomyak, *The principal term of the spectral asymptotics for "nonsmooth" elliptic problems*, Funktsional. Anal. i Prilozhen. **4** (1970), no. 4, 1–13; English transl. in Functional Anal. Appl. **4** (1970).

7. G. V. Rozenblyum, *The distribution of the discrete spectrum for singular differential operators*, Dokl. Akad. Nauk SSSR **202** (1972), 1012–1015; English transl. in Soviet Math. Dokl. **13** (1972).

8. J.-L. Lions and J. Peetre, *Sur une classe d'espaces d'interpolation*, Inst. Hautes Études Sci. Publ. Math. No. 19 (1964), 5–68.

9. Elias M. Stein and Guido Weiss, *Introduction to Fourier analysis on Euclidean spaces*, Princeton Univ. Press, Princeton, N. J., 1971.

10. Tord Holmstedt, *Interpolation of quasi-normed spaces*, Math. Scand. **26** (1970), 177–199.

11. G. E. Karadzhov, *An interpolation method of "means" for quasi-normed spaces*, Dokl. Akad. Nauk SSSR **209** (1973), 33–36; English transl. in Soviet Math. Dokl. **14** (1973).

12. M. Sh. Birman and M. Z. Solomyak, *Remarks on the nuclearity of integral operators and the boundedness of pseudodifferential operators*, Izv. Vyssh. Uchebn. Zaved. Matematika **1969**, no. 9 (88), 11–17. (Russian)

13. G. P. Kostometov, *Asymptotic behavior of the spectrum of integral operators with a singularity on the diagonal*, Mat. Sb. **94 (136)**(1974), 444–451; English transl. in Math. USSR Sb. **23** (1974).

14. Jaak Peetre, *Sur les espaces de Besov*, C.R. Acad. Sci. Paris Sér. A–B **264** (1967), A281–A283.

15. S. L. Sobolev, *The density of compactly supported functions in the space* $L_p^{(m)}(E_n)$, Sibirsk. Mat. Zh. **4** (1963), 673–682. (Russian)

16. O. V. Besov, a) *The continuation of functions in* L_p^l *and* W_p^l, Trudy Mat. Inst. Steklov. **89** (1967), 5–17; English transl. in Proc. Steklov Inst. Math. **89** (1967).

b) *On the density of functions with compact support in* $\mathcal{L}_{p,\theta}$ *and the extension of functions*, Trudy Mat. Inst. Steklov. **89** (1967), 18–30; English transl. in Proc. Steklov Inst. Math. **89** (1967).

17. Jaak Peetre, *On interpolation of* L_p *spaces with weight functions*, Acta Sci. Math. (Szeged) **28** (1967), 61–69.

18. G. N. Yakovlev, *Boundary properties of functions of class* $W_p^{(l)}$ *on regions with angular points*, Dokl. Akad. Nauk SSSR **140** (1961), 73–76; English transl. in Soviet Math. Dokl. **2** (1961).

19. V. A. Solonnikov, *A priori estimates for second-order parabolic equations*, Trudy Mat. Inst. Steklov. **70** (1964), 133–212; English transl. in Amer. Math. Soc. Transl. (2) **65** (1967).

20. John E. Gilbert, *Interpolation between weighted* L^p*-spaces*, Ark. Mat. **10** (1972), 235–249.

21. M. Sh. Birman and M. Z. Solomyak, *Spectral asymptotics of nonsmooth elliptic operators*. II, Trudy Moskov. Mat. Obshch. **28** (1973), 3–34; English transl. in Trans. Moscow Math. Soc. **28** (1973).

22. M. Sh. Birman and V. V. Borzov, *On the asymptotic formula for the discrete spectrum of certain singular differential operators*, Problemy Mat. Fiz., vyp. 5, Izdat. Leningrad Univ., Leningrad, 1971, pp. 24–38; English transl. in Topics in Math. Phys., no. 5, Plenum Press, New York, 1972.

23.* Jaak Peetre, *New thoughts on Besov spaces*, Math. Dept., Duke Univ., Durham, N.C., 1976.

Translated by J. PEETRE

*Added by translator.

Amer. Math. Soc. Transl.
(2) Vol. **127**, 1986

Asymptotic Behavior of Fredholm Determinants of Truncated Integral Operators with Matrix-Valued Kernels Depending on the Difference of the Arguments*

UDC 517.4

L. V. MIKAELYAN

Consider the integral equation

$$f(t) + \lambda \int_0^r K(t-s) f(s)\, ds = g(t) \qquad (0 \leqslant t \leqslant r, 0 < r < \infty), \tag{1}$$

where $K(t) = \|k_{ij}(t)\|_{i,j=1}^n$ $(-\infty < t < \infty)$ is a given matrix-valued function, $g(t) = \{g_j(t)\}_1^n$ is a given vector-valued function, and $f(t) = \{f_j(t)\}_1^n$ is an unknown vector-valued function. By applying the method proposed by Fredholm it is easy to see that this system is equivalent to the scalar equation

$$F(t) + \lambda \int_0^{nr} K_r(t,s) F(s)\, ds = G(t) \qquad (0 \leqslant t \leqslant nr); \tag{2}$$

here $F(t) = f_j(t-(j-1)r)$, $G(t) = g_j(t-(j-1)r)$ for $(j-1)r \leqslant t < jr$, and $K_r(t,s) = k_{ij}(t-(i-1)r - s + (j-1)r)$ for $(i-1)r \leqslant t < ir$, $(j-1)r \leqslant s < jr$ $(i, j = 1, 2, \ldots, n)$.

Under specific conditions the following Fredholm determinant is defined in the usual way for this equation:

$$D(K_r, \lambda) = 1 + \sum_{k=1}^{\infty} \frac{\lambda^k}{k!} \int_0^{nr} \cdots \int_0^{nr} K_r\begin{pmatrix} s_1 & \cdots & s_k \\ s_1 & \cdots & s_k \end{pmatrix} ds_1 \cdots ds_k, \tag{3}$$

where

$$K_r\begin{pmatrix} s_1 & \cdots & s_k \\ s_1 & \cdots & s_k \end{pmatrix} = \det\| K_r(s_i, s_j) \|_{i,j=1}^k$$

is the Fredholm symbol.

1980 *Mathematics Subject Classification*. Primary 45F05, 45M05; Secondary 15A15, 42A16, 47B35.
*Translation of Akad. Nauk Armyan. SSR Dokl. **68** (1979), 137–145; (MR **80m:** 45006).

Since the spectrum of the system (1) coincides with that of equation (2), it is natural to call $D(K_r, \lambda)$, denoted here by $D_r(K, \lambda)$, the *Fredholm determinant* of the system (1).

In this note we study the asymptotic behavior of $D_r(K, \lambda)$ as $r \to \infty$. At the same time we consider the continuous analogue for the behavior of the determinant of a Toeplitz block matrix $\|C_{i-j}\|_{i,j=1}^k$ as $k \to \infty$, where the C_j, $j = 0, \pm 1, \pm 2, \ldots$, are certain matrices of a fixed order n.

The latter problem was first studied by Szegö in the scalar case where the c_j are the Fourier coefficients of a positive function $p(t)$ [1]. Many articles were later devoted to the refinement and generalization of his results: M. G. Kreĭn [2], Hirschman [3], Devinatz [4], Geronimus [5], Golinskiĭ and Ibragimov [6], and others.

Bowers [7], Widom [8], and others considered the determination of the asymptotic behavior of $\det\|c_{i-j}\|_{i,j=1}^k$ as $k \to \infty$ in the case where the c_j are the Fourier coefficients of a matrix-valued function $c(t)$.

In the scalar case Kac [9] first discovered continuous analogues of Szegö's theorems by using probabilistic methods. Other results in this direction were obtained by Akhiezer [10], Hirschman [3], and Solev [11].

Some results on continuous analogues of Szegö's theorems for the matrix case were presented by Linnik in [12] and by the author in [13].

1°. We introduce the necessary notation and give some known and auxiliary propositions.

Let $L_1^{(n\times n)}$ be the Banach space of matrix-valued functions $\Phi(t) = \|\varphi_{ij}(t)\|_{i,j=1}^n$ with elements in $L_1(-\infty, \infty)$ and with the norm

$$|\Phi| = \int_{-\infty}^{+\infty} \sum_{i,j=1}^{n} |\varphi_{ij}(t)|\, dt.$$

It is not hard to verify that the set of all matrix-valued functions admitting the representation

$$\Phi(\lambda) = C_\Phi + \int_{-\infty}^{+\infty} \hat{\Phi}(t) e^{i\lambda t}\, dt, \tag{4}$$

where $C_\Phi = \|C_\Phi^{ij}\|_{i,j=1}^n$ is a constant matrix and $\hat{\Phi}(t) \in L_1^{(n\times n)}$, is a Banach algebra with respect to pointwise multiplication, under the norm here $|C_\Phi| = \sum_{i,j=1}^n |C_\Phi^{ij}|$. This algebra is denoted by $R^{(n\times n)}$. We define a subalgebra $R_+^{(n\times n)}$ ($R_-^{(n\times n)}$) of $R^{(n\times n)}$ as follows: $\Phi(\lambda) \in R_+^{(n\times n)}$ ($\Phi(\lambda) \in R_-^{(n\times n)}$) if $\hat{\Phi}(t) = 0$ for $t < 0$ ($t > 0$) in (4).

Let $L_2^{(n)}$ denote the Hilbert space of n-dimensional vector-valued functions on $(0, \infty)$ with inner product

$$(f, g) = \int_0^\infty g^*(t) f(t)\, dt.$$

If

$$K(\lambda) = I_n + \int_{-\infty}^{+\infty} \hat{K}(t) e^{i\lambda t}\, dt \in R^{(n\times n)},$$

then we define the following operators in $L_2^{(n)}$ associated with $K(\lambda)$:

$$T(K)f = f(t) + \int_0^\infty \hat{K}(t-s)f(s)\,ds \qquad (0 \leqslant t < \infty), \tag{5}$$

$$H(K)f = \int_0^\infty \hat{K}(t+s)f(s)\,ds \qquad (0 \leqslant t < \infty), \tag{6}$$

and also the operators

$$(P_r f)(t) = \begin{cases} f(t) & \text{if } t \in [0, r], \\ 0 & \text{if } t \in (r, \infty). \end{cases}$$

$$(Q_r f)(t) = \begin{cases} f(r-t) & \text{if } t \in [0, r], \\ 0 & \text{if } t \in (r, \infty). \end{cases}$$

It is easy to verify that

$$T(K_1K_2) - T(K_1)T(K_2) = H(K_1)H(\tilde{K}_2), \tag{7}$$

$$T_r(K_1K_2) - T_r(K_1)T_r(K_2) = P_rH(K_1)H(\tilde{K}_2)P_r + Q_rH(\tilde{K}_1)H(K_2)Q_r; \tag{8}$$

here $T_r(K) = P_rT(K)P_r$, and $\tilde{K}(\lambda) = I_n + \int_{-\infty}^{+\infty} \hat{K}(-t)e^{i\lambda t}\,dt$.

Consider the set $\mathbf{K}^{(n\times n)}$ consisting of the matrix-valued functions $\Phi(\lambda) \in R^{(n\times n)}$ such that the quantity[1]

$$\|\Phi\|_1 = |C_\Phi|_1 + \int_{-\infty}^{+\infty} |\Phi(t)|_1\,dt + \left(\int_{-\infty}^{+\infty} |t|\,|\Phi(t)|_1^2\,dt\right)^{1/2}$$

is finite. If this quantity is taken as a norm, then it follows from [2] that $\mathbf{K}^{(n\times n)}$ is a Banach algebra.

It is known that if $K(\lambda) \in \mathbf{K}^{(n\times n)}$, then the operators $H(K)$ and $H(\tilde{K})$ are Hilbert-Schmidt operators [14].

We mention two lemmas that are useful in what follows.

Consider the vector integral equation

$$f(t) + \lambda\int_0^r K(t,s)f(s)\,ds = g(t). \tag{9}$$

For this equation we define the Fredholm determinant $D_r(K,\lambda)$ as in (3), taking

$$K_r(t,s) = k_{ij}(t-(i-1)r, s-(j-1)r)$$
$$((i-1)r \leqslant t < ir, \quad (j-1)r \leqslant s < jr, \quad i,j = 1,2,\ldots,n),$$

where $K(t,s) = \|k_{ij}(t,s)\|_{i,j=1}^n$.

LEMMA 1. *If the matrix-valued function* $K(t,s)$ *equals* 0 *for* $t > s$ *or for* $t < s$ *in* (9), *then*

$$D_r(K,\lambda) = \exp\left[\lambda\int_0^r \operatorname{tr} K(s,s)\,ds\right].$$

In addition to this lemma we need

LEMMA 2. *If the matrix-valued function*

$$K(\lambda) = I_n + \int_{-\infty}^{+\infty} K(t)e^{i\lambda t}\,dt \in R^{(n\times n)}$$

---

[1] If $A = \|a_{ij}\|_{i,j=1}^n$, then $|A|_1 = (\sum_{i,j=1}^n |a_{ij}|^2)^{1/2}$.

admits a factorization

$$I_n + \int_{-\infty}^{+\infty} \hat{K}(t)e^{i\lambda t}\,dt = \left(I_n + \int_0^{\infty} \hat{K}(t)e^{i\lambda t}\,dt\right)\left(I^n + \int_{-\infty}^{0} \hat{K}_-(t)e^{i\lambda t}\,dt\right),$$

where the matrix-valued functions $\hat{K}_+(t)$ *and* $\hat{K}_-(t)$ *are continuous at zero, then*

$$\operatorname{tr}[\hat{K}_+(0) + \hat{K}_-(0)] = \lim_{T\to\infty} \frac{1}{2\pi}\int_{-T}^{T}\left(1 - \frac{|\lambda|}{T}\right)\ln\det K(\lambda)\,d\lambda.$$

2°. Suppose that

$$K(\lambda) = I_n + \int_{-\infty}^{+\infty} \hat{K}(t)e^{i\lambda t}\,dt \in R^{(n\times n)}$$

is invertible in this algebra. Its inverse is denoted by

$$K^{-1}(\lambda) = I_n + \int_{-\infty}^{+\infty} \hat{K}^0(t)e^{i\lambda t}\,dt \in R^{(n\times n)}.$$

The operators $T(K^{-1})$ and $H(K^{-1})$ are defined by (5) and (6), with the matrix-valued function $\hat{K}(t)$ replaced by $\hat{K}^0(t)$ in them. Let $D_r(K) = D_r(\hat{K}, 1)$.

THEOREM 1. *If*

$$K(\lambda) = I_n + \int_{-\infty}^{+\infty} \hat{K}(t)e^{i\lambda t}\,dt \in \mathbf{K}^{(n\times n)}$$

admits a factorization

$$K(\lambda) = K_+(\lambda)K_-(\lambda) = \left(I_n + \int_0^{\infty} \hat{K}_+(t)e^{i\lambda t}\,dt\right)\left(I_n + \int_{-\infty}^{0} \hat{K}_-(t)e^{i\lambda t}\,dt\right), \tag{10}$$

with $K_{\pm 1}^{\pm 1}(\lambda) \in \mathbf{K}^{(n\times n)}$ *and the matrix-valued functions* $\hat{K}_{\pm}(t)$ *continuous at zero and bounded on each finite interval, then*

$$\lim_{r\to\infty} \frac{D_r(K)}{[G(K)]^r} = \det[T(K)T(K^{-1})],$$

where

$$G(K) = \exp\left\{\lim_{T\to\infty} \frac{1}{2\pi}\int_{-T}^{T}\left(1 - \frac{|\lambda|}{T}\right)\ln\det K(\lambda)\,d\lambda\right\}.$$

PROOF. From (10) it follows that

$$K(\lambda)K_-^{-1}(\lambda) = K_+(\lambda).$$

Applying (8) and taking account of the fact that $H(K_-^{-1}) = 0$, we have

$$T_r(K_+) - T_r(K)T_r(K_-^{-1}) = P_rH(K)H(\tilde{K}_-^{-1})P_r;$$

hence

$$T_r(K_+)T_r(K_+^{-1}) - T_r(K)T_r(K_-^{-1})T_r(K_+^{-1}) = P_rH(K)H(\tilde{K}_-^{-1})P_rT_r(K_+^{-1}). \tag{11}$$

From (8) it is not hard to see that

$$T_r(K_+)T_r(K_+^{-1}) = P_r, \qquad T_r(K_-)T_r(K_-^{-1}) = P_r. \tag{12}$$

Therefore, (11) takes the following form:

$$T_r(K)T_r(K_-^{-1})T_r(K_+^{-1}) = P_r - P_rH(K)H(\tilde{K}_-^{-1})P_rT_r(K_+^{-1}). \tag{13}$$

As mentioned above, under the conditions of the theorem $H(K)$ and $H(\tilde{K}_-^{-1})$ are Hilbert-Schmidt operators. Hence, the second term on the right-hand side of (13) is a nuclear operator. In this case [14] the Fredholm determinant of the right-hand side of (13) coincides with the expression

$$\det\left[P_r - P_rH(K)H(K_-^{-1})P_rT_r(K_+^{-1})\right] = \prod_j (1 - \lambda_j(r)),$$

where the $\lambda_j(r)$ are the eigenvalues of the operator $P_rH(K)H(\tilde{K}_-^{-1})P_rT_r(K_+^{-1})$.

Taking account of the fact that the Fredholm determinant of the integral operator on the left-hand side of (13) at the point $\lambda = 1$ equals the quantity $D_r(K)D_r(K_-^{-1})D_r(K_+^{-1})$, we have

$$D_r(K)D_r(K_-^{-1})D_r(K_+^{-1}) = \det\left[P_r - P_rH(K)H(\tilde{K}_-^{-1})P_rT_r(K_+^{-1})\right]. \tag{14}$$

Since $D_r(K_\pm^{-1}) = [D_r(K_\pm)]^{-1}$, (14) can be rewritten in the form

$$\frac{D_r(K)}{D_r(K_+)D_r(K_-)} = \det\left[P_r - P_rH(K)H(\tilde{K}_-^{-1})P_rT_r(K_+^{-1})\right]. \tag{15}$$

By Lemma 1,

$$D_r(K_\pm) = \exp\left\{\int_0^r \operatorname{tr}\hat{K}_\pm(s-s)\,ds\right\} = \exp\left[r\cdot\operatorname{tr}\hat{K}_\pm(0)\right].$$

Therefore, it can be asserted by Lemma 2 that the denominator of the left-hand side of (15) equals $[G(K)]^r$.

Hence, (15) can be written in the form

$$\frac{D_r(K)}{[G(K)]^r} = \det\left[P_r - P_rH(K)H(\tilde{K}_-^{-1})P_rT_r(K_+^{-1})\right].$$

Obviously, $P_r \to I$ in the strong sense as $r \to \infty$, where I is the identity operator in $L_2^{(n)}$. Therefore,

$$P_rH(K)H(\tilde{K}_-^{-1})P_rT_r(K_+^{-1}) \to H(K)H(\tilde{K}_-^{-1})T(K_+^{-1})$$

in the nuclear metric. Thus,

$$\lim_{r\to\infty}\frac{D_r(K)}{[G(K)]^r} = \det\left[I - H(K)H(\tilde{K}_-^{-1})T(K_+^{-1})\right].$$

It is not hard to verify that

$$H(K_1K_2) = H(K_1)T(\tilde{K}_2) + T(K_1)H(K_2).$$

From this, substituting $K_1 = \tilde{K}_-^{-1}$, $K_2 = \tilde{K}_+^{-1}$ and using the fact that $H(\tilde{K}_+^{-1}) = 0$, we get

$$H(\tilde{K}^{-1}) = H(\tilde{K}_-^{-1})T(K_+^{-1}).$$

Consequently,

$$I - H(K)H(K_-^{-1})T(K_+^{-1}) = I - H(K)H(\tilde{K}^{-1}) = T(K)T(K^{-1}).$$

The theorem is proved.

THEOREM 2. *If*

$$K(\lambda) = I_n + \int_{-\infty}^{+\infty} \hat{K}(t) e^{i\lambda t}\, dt \in \mathbf{K}^{(n\times n)},$$

and, moreover, $\det K(\lambda) \neq 0$ $(-\infty < \lambda < \infty)$, $V_{-\infty}^{+\infty} \arg \det K(\lambda) = 0$ *and* $\hat{K}(t)$ *is continuous at zero and bounded on each finite interval, then*

$$\lim_{r\to\infty} \frac{D_r(K)}{[G(K)]^r} = \det[T(K)T(K^{-1})].$$

PROOF. Under the conditions of the theorem, there exists a matrix-valued function

$$K_1(\lambda) = \int_{-\infty}^{+\infty} \hat{K}_1(t) e^{i\lambda t}\, dt \in \mathbf{K}^{(n\times n)}$$

such that the matrix-valued function $K(\lambda) + \varepsilon K_1(\lambda)$ admits the factorization required in Theorem 1 for all sufficiently small $\varepsilon \neq 0$. Therefore, for such ε we have

$$\frac{D_r(K + \varepsilon K_1)}{[G(K + \varepsilon K_1)]^r} = \det\left[T(K + \varepsilon K_1)T\left((K + \varepsilon K_1)^{-1}\right)\right].$$

Since all the quantities in this formula depend analytically on ε, we can assert that it is also true for $\varepsilon = 0$.

When $K(\lambda)$ is a scalar function, it is possible to calculate $\det[T(K)T(K^{-1})]$. More precisely, we have

THEOREM 3. *If*

$$K(\lambda) = \int_{-\infty}^{+\infty} \hat{K}(t) e^{i\lambda t}\, dt \in \mathbf{K}^{(n\times n)},$$

and, moreover, $K(\lambda) \neq 0$ $(-\infty < \lambda < \infty)$, $V_{-\infty}^{+\infty} \arg K(\lambda) = 0$, *and* $\hat{K}(t)$ *is continuous at zero and bounded on each finite interval, then*

$$\lim_{r\to\infty} \frac{D_r(K)}{[G(K)]^r} = \exp\left\{\int_0^{\infty} t l(t) l(-t)\, dt\right\},$$

where

$$G(K) = \exp\left\{\lim_{T\to\infty} \frac{1}{2\pi} \int_{-T}^{T} \left(1 - \frac{|\lambda|}{T}\right) \ln K(\lambda)\, d\lambda\right\},$$

$$\ln K(\lambda) = \int_{-\infty}^{+\infty} l(t) e^{i\lambda t}\, dt.$$

The formulation of the problem is due to M. G. Kreĭn. The results were discussed with him and with V. M. Adamyan, and the author sincerely thanks them for this.

REFERENCES

1. G. Szegö, *Ein Grenzwertsatze über die Toeplitz'schen Determinanten einer reellen positiven Funktion*, Math. Ann. **76** (1915), 490–503.

2. M. G. Kreĭn, *On some new Banach algebras and Wiener-Levi type theorems for Fourier series and integrals*, Mat. Issled. **1** (1966), no. 1, 82–109; English transl. in Amer. Math. Soc. Transl. (2) **93** (1970).

3. I. I. Hirschman, Jr., *On a formula of Kac and Achiezer*, J. Math. and Mech. **16** (1966), 167–196.

4. A. Devinatz, *The strong Szegö limit theorem*, Illinois J. Math. **11** (1967), 160–175.

5. Ya. L. Geronimus, *On a problem of Szegö, Kac, Baxter and Hirschman*, Izv. Akad. Nauk SSSR Ser. Mat. **31** (1967), 289–304; English transl. in Math. USSR Izv. **1** (1967).

6. B. L. Golinskiĭ and I. A. Ibragimov, *On Szegö's limit theorem*, Izv. Akad. Nauk SSSR Ser. Mat. **35** (1971), 408–427; English transl. in Math. USSR Izv. **5** (1971).

7. Newton Lynn Bowers, Jr., *Toeplitz forms associated with matrix valued functions*, Ph.D. Thesis, Univ. of Minnesota, Minneapolis, Minn., 1965.

8. Harold Widom, *Asymptotic behavior of block Toeplitz matrices and determinants*. I, II, Advances in Math. **13** (1974), 284–322; ibid. **21** (1976), 1–29.

9. M. Kac, *Toeplitz matrices, translation kernels and a related problem in probability theory*, Duke Math. J. **21** (1954), 501–509.

10. N. I. Akhiezer, *A continual analog of some theorems on Toeplitz matrices*, Ukrain. Mat. Zh. **16** (1964), 445–462; English transl. in Amer. Math. Soc. Transl. (2) **50** (1966).

11. V. N. Solev, *The continual analog of a theorem of Szegö*, Zap. Nauchn. Sem. Leningrad. Otdel. Mat. Inst. Steklov. (LOMI) **39** (1974), 104–109; English transl. in J. Soviet Math. **8** (1977), no. 1.

12. I. Yu. Linnik, *A higher-dimensional generalization of a theorem on Toeplitz operators*, Mat. Zametki **14** (1973), 895–900; English transl. in Math. Notes **14** (1973).

13. L. V. Mikaelyan, *The multidimensional continual analog of a theorem of Szegö*, Izv. Akad. Nauk Armyan. SSR Ser. Mat. **11** (1976) , 275–286. (Russian)

14. I. Ts. Gokhberg [Israel Gohberg] and M. G. Kreĭn, *Introduction to the theory of linear nonselfadjoint operators in Hilbert space*, "Nauka", Moscow, 1965; English transl., Amer. Math. Soc., Providence, R.I., 1969.

Translated by H. H. MCFADEN

Amer. Math. Soc. Transl.
(2) Vol. **127**, 1986

Optimization of Algorithms for the Solution of Singular Integral Equations with Respect to Their Realization Time on a Computer*

V. V. IVANOV AND G. N. KOTENKOVA

§1. Statement of the problem

A large class of problems of computational mathematics reduce to the solution of operator equations of the form

$$Ax = y, \tag{1}$$

where A is an operator with domain $D(A)$ and range $R(A)$, y is a given element, and x is the unknown. More often than not, equation (1) cannot be solved exactly, and it is therefore necessary to develop methods to solve it approximately. One of the most interesting methods for the approximate solution of regular integral equations, which is optimal with respect to the order of the number of operations, is considered in [1].([1]) In [2]–[4] primary consideration is given to the question of solutions of optimal accuracy for general operator and singular integral equations. It has been noted that in the case of singular integral equations one can construct combined algorithms, using both direct and iterative approaches, which will also be optimal with respect to the order of the number of arithmetical operations.

In this article we investigate a combined method of solving a linear equation of the form (1), which is optimal with respect to the order of the number of operations under a given restriction on the error of the method or is a method of maximum (with respect to order) accuracy for a given bound on the number of

1980 *Mathematics Subject Classification*. Primary 65J10, 45L10, 65R20; Secondary 45E05, 65J15.

*Translation of Differential and Integral Equations. Boundary Value Problems (I. N. Vekua Memorial Coll; R. A. Kordzadze, editor), Izdat. Tbiliss. Gos. Univ., Tbibsi, 1979, pp. 113–126; MR **81k:** 65146.

([1])Similar results were obtained almost simultaneously by V. I. Lebedev, S. I. Tsatselkin and others.

operations, and we consider in more detail special cases of singular integral equations. Some approaches to the evaluation of the number of operations in the solution of nonlinear operator equations are also indicated.

§2. On the least squares method

Let $D(A) \subset X$ and $R(A) \subset Y$, where X and Y are Hilbert spaces. According to the least squares method, an approximate solution of the linear equation (1) will be sought in the form

$$x_n = \sum_{k=1}^{n} \alpha_k e_k, \qquad e_k \in D(A), \tag{2}$$

and the unknown values α_k^* will be found (taking into account the regularization method [5], in the case of approximate initial data) from the condition of minimum of the quadratic function

$$\Delta_n(P_n) = I^{(\alpha)}(x_n) \equiv \left\| y - \sum_{k=1}^{n} \alpha_k y_k \right\|^2 + \alpha \left\| \sum_{1}^{n} \alpha_k e_k \right\|^2,$$

$$\alpha \geqslant 0, \; y_k = Ae_k, \; P_n = (\alpha_1, \alpha_2 \cdots \alpha_n). \tag{3}$$

The smallest value of $\Delta_n(P_n)$ is always attained at at least one point $P_n^*(\alpha_1^*, \alpha_2^*, \ldots, \alpha_n^*)$, which is equivalent to the solvability of the normal system of algebraic equations resulting from the necessary conditions for a minimum:

$$\partial \Delta_n(P_n)/\partial \alpha_j = 0, \qquad j = 1, \ldots, n,$$

$$\sum_{k=1}^{n} \alpha_k \left[\operatorname{Re}(y_k, y_j) + \alpha \operatorname{Re}(e_k, e_j) \right] = \operatorname{Re}(y, y_j), \qquad j = 1, \ldots, n. \tag{4}$$

We assume here that all the α_k are real, which can always be achieved by introducing, if necessary, along with the system e_k, the system ie_k, $i = \sqrt{-1}$. We shall also assume that $\{e_k\}_1^\infty$ is an orthonormal system. Then

$$\Delta_n(P_n) = I^{(\alpha)}(x_n) \equiv \left\| y - \sum_{1}^{n} \alpha_k y_k \right\|^2 + \alpha \sum_{1}^{n} \alpha_k^2, \qquad y_k = Ae_k. \tag{5}$$

Let us assume that $e_1, \ldots, e_n$ are the first n elements of an A-complete system in $\overline{R(A)}$: any element of $\overline{R(A)}$ can be approximated arbitrarily closely by linear combinations of the Ae_k. Then, as shown in [6], under very weak constraints

$$\lim_{n \to \infty} \Delta_n(P_n^*) = I^{(\alpha)}(x^{*(\alpha)}) = \inf_{x \in D(A)} I^{(\alpha)}(x),$$

$$I^{(\alpha)} = \| y - Ax \|^2 + \alpha \| x \|^2 \tag{6}$$

and

$$\| x_n^* - x^{*(\alpha)} \|^2 \leqslant (1/\alpha) \left[I^{(\alpha)}(x_n^*) - I^{(\alpha)}(x^{*(\alpha)}) \right] \to 0, \qquad \alpha < 0, \; n \to \infty,$$

$$x_n^* = \sum_{1}^{n} \alpha_k^* e_k. \tag{7}$$

In general, the rate of convergence of $\|x_n^* - x^{*(\alpha)}\|$ to zero is arbitrary.

If the operator A is normally solvable, then for $\alpha = 0$

$$\inf_{\substack{I(x^0)=\inf I(x)\\ x\in D(A)}} \left\| x_n^* - x_0 \right\| \leqslant \|A^{-1}\| \, \|Ax_n^* - Ax^*\| \leqslant \|A^{-1}\| \, \|A\| \rho_n(x^*),$$

$$J(x) = \|y - Ax\|, \qquad \rho_n(x^*) = \inf_{\{\alpha_k\}_1^n} \left\| x^* - \sum_1^n \alpha_k e_k \right\|, \tag{8}$$

where A^{-1} is a right inverse of A, i.e. $D(A^{-1}) = R(A)$ and $R(A^{-1}) = D(A) \dot{-} L(A)$, $L(A)$ being the null space of the operator A; and $\rho_n(x^*)$ is the value of the best approximation to x^* (x^* being that element from x^0 which is orthogonal to $L(A)$).

An important role in many relations is played by the quantity

$$m_n = \inf_{\|\sum_1^n \alpha_k\| = 1} \left\| \sum_1^n \alpha_k y_k \right\|, \qquad y_k = Ae_k. \tag{9}$$

If we denote by λ_n the magnitude of the smallest eigenvalue of the matrix of the system (4) with $\alpha = 0$, then

$$m_n = \sqrt{\lambda_n}\,. \tag{10}$$

The following estimate holds:

$$\left\| x_n^* - x^* \right\| \leqslant \left(1 + 2\|A\|/m_n\right) \rho_n(x^*). \tag{11}$$

If an estimate of the form

$$\rho_n(x^*) \leqslant \varepsilon_n \|x^*\| \tag{12}$$

is obtained, then for small enough ε_n

$$\varepsilon\left(1 + 2\|A\|/m_n\right) < 1, \tag{13}$$

and from (11) we get

$$\left\| x_n^* - x^* \right\| \leqslant \varepsilon_n \left\| x_n^* \right\| \frac{\left(1 + 2\|A\|/m_n\right)}{1 - \varepsilon_n\left(1 + 2\|A\|/m_n\right)}. \tag{14}$$

The estimate (12) can be obtained from the properties of specific classes of equations.

It is easy to see that

$$m_n \geqslant m = \inf_x \frac{\|Ax\|}{\|x\|}. \tag{15}$$

If (in the case of functional spaces, when x^* is sufficiently smooth)

$$\varepsilon_n \asymp n^{-s}, \qquad m > 0, \tag{16}$$

then it follows from (14) that

$$\left\| x_n^* - x_* \right\| = \asymp n^{-s}. \tag{17}$$

However, the case where $m > 0$ corresponds to well-posed problems. Let $m = 0$. Then one can only assert the convergence of the method under consideration with respect to residuals:

$$I(x_n) = \| Ax_n - y \| \to 0, \qquad n \to \infty, \tag{18}$$

where $y' = y$ if $y \in \overline{R(A)}$, and $y' - y \perp \overline{R(A)}$ in general; and also the rate of convergence can be different and in a number of special cases can be of power type (17).

If one must find an approximate solution converging to an element x^*, $\|Ax^* - y\| = \inf_{x \in D(A)} \|Ax - y\|$, one can modify the above scheme of the least squares method by searching for the minimum of the quadratic function

$$\delta_n(P_n) = \left\| x^* - \sum_1^n \alpha_k A^* e_k \right\|^2 + \alpha \sum_{k=1}^n \alpha_k^2. \tag{19}$$

The condition for a minimum yields the following normal system of algebraic equations:

$$\sum_{k=1}^n \alpha_k \left[\operatorname{Re}\left(y_k^*, y_j' \right) + \alpha \delta_{kj} \right] = \operatorname{Re}(y, e_j), \qquad j = 1, \dots, n,$$

$$y_k^* = A^* e_k, \qquad \delta_{kj} = \begin{cases} 1, & k = j, \\ 0, & k \neq j. \end{cases} \tag{20}$$

If $\{A^* e_k\}_1^\infty$ is a complete system in $\overline{D(A)}$, then

$$\delta_n(P_n) \to 0, \qquad n \to \infty, \tag{21}$$

and then the convergence rate in a number of special cases will also have the form (17).

In the case of approximate initial data, for the choice of α corresponding to an assigned accuracy, see [5] and [6].

§3. Main results

In order to reduce the number of arithmetic operations in the systems (4) and (20), which in matrix form read

$$BP_n = b, \qquad B = \{ b_{ij} \}, \tag{22}$$

we will approximate the matrix B by a $(2k + 1)$-diagonal matrix B_k whose elements are

$$\begin{cases} b_{ij}^{(k)} = b_{ij}, & |j - i| \leqslant k, \\ b_{ij}^{(k)} = 0, & |j - i| > k, 0 \leqslant k \leqslant n. \end{cases} \tag{23}$$

Let us assume that

$$\| B_l - B_m \|_3 \leqslant c/m^s, \quad n \geqslant l \geqslant m \qquad (B_n \equiv B), \tag{24}$$

where c is a known constant, independent of n. This condition is fulfilled by a very large class of operators A and systems $\{l_k\}_1^n$. Indeed, in the case of a Fredholm operator

$$A_1 x \equiv \int_\gamma k(t,\tau)x(\tau)\,d\tau, \tag{25}$$

where γ is a circle of unit radius centered at the origin, and for the system

$$\{t^k\}_{-n_1}^{n_1},\ t = e^{i\theta}, \qquad -\pi \leqslant \theta \leqslant \pi\ (n = 2n_1 + 1), \tag{26}$$

we have, for $|k| \leqslant m/2$,

$$\sum_{l \geqslant |k-j| > m} \left[\operatorname{Re}(y_k, y_j)\right]^2 = \frac{1}{2\pi} \sum_{l \geqslant |k-j| > m} \left[\int_\gamma \int_\gamma \bar{k}(t,\tau)\tau^k d\tau \cdot t^{-j}\frac{dk}{t}\right]^2 \leqslant \frac{c_1^2}{m^{2s}},$$

$$\bar{k}(t,\tau) = \int k(\tau,v)\bar{k}(t,v)\,dv, \tag{27}$$

and, for $|k| > m/2$,

$$\sum_{l \geqslant |k-j| > m} \left[\operatorname{Re}(y_k, y_j)\right]^2 \leqslant \sum_{l \geqslant |k-j| > m} \frac{1}{|k, j|_2^2}$$

$$\leqslant \frac{4}{m^{2s}}\left(\sum_1^n \frac{1}{|j|^{2s}}\right) \leqslant \frac{c_2^2}{m^{2s}}, \tag{28}$$

provided $\bar{k}(t,\tau)$ has a derivative $\partial^{2s}\bar{k}(t,\tau)/\partial t^s \partial \tau^s$ bounded in absolute value [7], for which it is sufficient that $|k_t^{(s)}(t,\tau)|^2$ be uniformly integrable with respect to τ. A similar estimate can easily be obtained also in the case of the system (20), for which the sufficient condition will be the uniform integrability of the function $|k_\tau^{(s)}(t,\tau)|^2$ with respect to t.

In the case of a non-Fredholm operator, e.g. for the operator

$$A_2 x \equiv \int_0^\infty k(t-\tau)x(\tau)\,d\tau, \tag{29}$$

and the system

$$\psi_k(t) = \begin{cases} e^{-t}P_k(t), & t > 0, \\ 0, & t < 0,\ k \geqslant 0; \end{cases}$$
$$\psi_k(t) = \begin{cases} 0, & t > 0, \\ e^t P_{-k-1}(-t), & t < 0,\ k < 0, \end{cases} \tag{30}$$

where the $P_k(t)$ are the Laguerre polynomials

$$\int_0^\infty e^{-2t}P_k(t)P_j(t)\,dt = \delta_{kj} = \begin{cases} 1, & k = j, \\ 0, & k \neq j, \end{cases} \tag{31}$$

we find that

$$\|B_l - B_m\| = \sup_{\|P_n\|=1} ((B_l - B_m)P_n, P_n)$$

$$= \sup_{\|X_n\|=1} \left(\left(|K|_l^2 - |K|_m^2\right)X_n, X_n\right) \leqslant \frac{c_3}{m^s}, \tag{32}$$

where

$$|K|_r^2 = \sum_{p=-r}^{r} \left(|K|^2, \Psi_p\right)\Psi_p, \qquad r = l, m, \quad P_n = (\alpha_1, \ldots, \alpha_n),$$

$$X_n = \sum_{p=1}^{n} \alpha_p \Psi_p$$

under the condition that

$$\left| K\left(i\frac{1+t}{1-t}\right)\frac{1}{1-t}\right|^2$$

has an sth derivative, Hölder continuous with index $\varepsilon > 0$. Here capital letters denote the Fourier transforms of the corresponding lower case letters; in particular

$$\Psi_k(\omega) = \frac{1}{\sqrt{2\pi}} \int_{-\infty}^{\infty} \psi_k(t) e^{it\omega}\, dt = \begin{cases} \dfrac{1}{\sqrt{\pi}} \left(\dfrac{\omega - i}{\omega + i}\right)^k \dfrac{i}{\omega + i}, & k \geqslant 0, \\ \dfrac{1}{\sqrt{\pi}} \left(\dfrac{\omega + i}{\omega - i}\right)^k \dfrac{i}{\omega - i}, & k < 0. \end{cases} \tag{33}$$

We note that if instead of the matrix $B = \{|K|_n^2 \Psi_k, \Psi_j\}_{k,j=1}^n$ we take the matrix $\tilde{B} = \{|\tilde{K}|_n^2 \Psi_k, \Psi_j\}$, where $||\tilde{K}|_n^2 - |K|^2| \leqslant c_4/n^s$, we can set $\varepsilon = 0$ in the above indicated condition (adding the condition of boundedness of the sth derivative). One can similarly show that an inequality of the form (24) also holds under corresponding conditions for operators A consisting of linear combinations of the identity, Fredholm and singular operators, their adjoints and a number of others.

Let us choose k_0, independent of n, such that

$$\left\| B_{k_0}^{-1}\left(B_{k_0} - B_l\right) \right\|_3 \leqslant \left\| B_{k_0}^{-1} \right\|_3 c/k_0^s = q_0 < 1. \tag{34}$$

For $\alpha > 0$ and for $\alpha = 0$ and $m > 0$, such a choice is always possible.

We introduce the notation

$$\begin{aligned} r_0 &= B_{k_0}^{-1} b, \\ z_0^{(\kappa)} &= \left[B_{k_0}^{-1}\left(B_{k_0} - B\right)\right] z_0, \qquad \kappa = 1, \ldots, r \end{aligned} \tag{35}$$

and take

$$P_n^{(r)} = z_0 + z_0^{(1)} + \cdots + z_0^{(r)}. \tag{36}$$

Then

$$\left\| P_n - P_n^{(r)} \right\| \leqslant \frac{q_0^{r+1}}{1 - q_0} \| z_0 \|. \tag{37}$$

We now set

$$z_\kappa = B_{k_0}^{-1} y_\kappa, \quad y_\kappa = \left(B_{k_0} - B_{k_\kappa}\right) z_{\kappa - 1}, \quad q_\kappa = \left\| B_{k_0}^{-1} \right\| c/k_\kappa^s, \qquad \kappa = 1, \ldots, r, \tag{38}$$

where

$$k_\kappa = [q^\kappa n] + 1, \qquad q < 1, \quad \kappa = 1, \dots, r. \tag{39}$$

We seek an approximate solution of (22) in the form

$$\tilde{P}_n^{(r)} = z_0 + B_{k_0}^{-1}(y_1 + y_2 + \cdots + y_r), \qquad \tilde{P}_n^{(r)} = \left(\tilde{\alpha}_1^{(r)}, \dots, \tilde{\alpha}_n^{(r)}\right). \tag{40}$$

We estimate the magnitude of the errors $\|z_\kappa - z_0^{(\kappa)}\|$:

$$\begin{aligned}
&\left\|z_1 - z_0^{(1)}\right\| \leqslant q_1\|z_0\|,\\
&\left\|z_2 - z_0^{(2)}\right\| = \left\|B_{k_0}^{-1}(B_{k_0} - B_{k_2})z_1 - B_{k_0}^{-1}(B_{k_0} - B)z_0^{(1)}\right\|\\
&\qquad \leqslant q_0\left\|z_1 - z_0^{(1)}\right\| + q_2\left\|z_0^{(1)}\right\|\\
&\qquad \leqslant q_0 q_1\|z_0\| + q_0 q_2\|z_0\| = q_0(q_1 + q_2)\|z_0\|,\\
&\left\|z_3 - z_0^{(3)}\right\| = \left\|B_{k_0}^{-1}(B_{k_0} - B_{k_3})\left(z_2 - z_0^{(2)}\right) + B_{k_0}^{-1}(B - B_{k_3})z_0^{(2)}\right\|\\
&\qquad \leqslant q_0^2(q_1 + q_2 + q_3)\|z_0\|.
\end{aligned}$$

and so forth. Therefore,

$$\left\|z_\kappa - z_0^{(\kappa)}\right\| \leqslant q_0^{\kappa-1}(q_1 + q_2 + \cdots + q_\kappa)\|z_0\|. \tag{41}$$

Hence, from (38) and (39) we get

$$\left\|z_\kappa - z_0^{(\kappa)}\right\| \leqslant \frac{C\left\|B_{k_0}^{-1}\right\|q_0^{\kappa-1}\|z_0\|}{n^3}\left(\frac{1}{q^s} + \cdots + \frac{1}{q^{s\kappa}}\right). \tag{42}$$

The error for the solution of (22) will be

$$\begin{aligned}
&\left\|P_n - \tilde{P}_n^{(r)}\right\| \leqslant \left\|P_n - P_n^{(r)}\right\| + \left\|P_n^{(r)} - \tilde{P}_n^{(r)}\right\|\\
&\qquad \leqslant \frac{q_0^{r+1}}{1 - q_0} + \|z_0\| + \frac{C\left\|B_{k_0}^{-1}\right\|\,\|z_0\|}{n^s}\sum_{\kappa=1}^{r} q_0^{\kappa-1}\left(\frac{1}{q^s} + \cdots + \frac{1}{q^{s\kappa}}\right)\\
&\qquad \leqslant \frac{q_0^{r+1}}{1 - q_0}\|z_0\| + \frac{C\left\|B_{k_0}^{-1}\right\|\,\|z_0\|}{n^s}\sum_{\kappa=1}^{r}\left(\frac{q_0}{q^s}\right)^{\kappa-1} \cdot \frac{(1/q^s)^2}{1/q^2 - 1}.
\end{aligned} \tag{43}$$

Choosing $q < 1$ from the condition

$$q_0/q^s = q_1 < 1, \tag{44}$$

we find that the error

$$\left\|P_n - \tilde{P}_n^{(r)}\right\| \leqslant \frac{q_0^{r+1}}{1 - q_0}\|z_0\| + \frac{C\left\|B_{k_0}^{-1}\right\|\,\|z_0\|}{n^s}\,\frac{q_1^2}{q_0(q_1 - q_0)(1 - q_1)}. \tag{45}$$

Finally, let us choose r from the condition

$$q_0^r = D/n^s, \qquad D \geqslant 1. \tag{46}$$

Then

$$\left\|P_n - \tilde{P}_n^{(r)}\right\| \leqslant \left(\frac{Dq_0}{1 - q_0} + \frac{C\left\|B_{k_0}^{-1}\right\|q_1^2}{q_0(q_1 - q_0)(1 - q_1)}\right)\left\|z^0\right\|\frac{1}{n^s} = \frac{E}{n^s}. \tag{47}$$

The number of additions and multiplications in the determination of $\tilde{P}_n^{(r)}$ (see (41)) will be

$$N \asymp rk_0^2 n + n(k_1 + k_2 + \cdots + k_r) < rk_0^2 n + (nr + n^2/(1 - q)) \leqslant rk_0^2 n + n\frac{s}{\ln(1/q_0)}\ln n + n^2\frac{q}{1 - q} \asymp n^2. \tag{48}$$

We summarize these results in the form of the following theorem.

THEOREM 1. *If the conditions* (17), (24) *and* (34) *are fulfilled, then*

$$\left\| x^* - \sum_{k=1}^{n} \tilde{\alpha}_k^{(r)} e_k \right\| \asymp \frac{1}{N^{s/2}}, \tag{49}$$

where N is the number of arithmetic operations of the algorithm (41) *for the determination of the unknown coefficients $\tilde{\alpha}_k^{(r)}$, $k = 1, \ldots, n$.*

Any algorithm using $O(n^2)$ distinct data from the $n(n + 1)/2$ data in $I_n(B, b)$ will have $N = O(n^2)$ binary and unary operations, since at least one operation will be performed over each argument of the algorithm. Let us show that for the class of problems containing two problems of the form (1), for which y can take two values

$$y_1, A^*y_1 = \sum_1^n u_k e_k; \qquad y_2, (A^*y_2, e_k) = u_k, \quad k = 1, \ldots, n, \qquad \left\| A^*y_2 - \sum_{k=1}^{n} u_k e_k \right\| = \frac{1}{n^s}; \qquad (A^*Ax, x) \leqslant M(x, x), \tag{50}$$

there is no algorithm using data from $I_n(B, b)$ by means of which one can obtain an approximate solution closer to the exact solution by less than $O(1/n^s)$.

Indeed, any such algorithm uses no more than n values $(A^*y, e_k) = u_k$. Let us take two solutions, x_1 and x_2:

$$A^*Ax_1 = A^*y_1 = \sum_{k=1}^{n} u_k, e_k, \qquad A^*Ax_2 = A^*y_2. \tag{51}$$

We have

$$\| A^*A(x_1 - x_2) \| = \left\| A^*y_2 - \sum_{k=1}^{n} u_k e_k \right\| = \frac{1}{n^s}, \tag{52}$$

whence

$$\| x_1 - x_2 \| \geqslant 1/Mn^s. \tag{53}$$

In order to obtain approximate values of the two solutions considered, the algorithm uses one and the same information. Consequently, the approximate solution to at least one of the solutions x_1 or x_2 cannot be closer than $1/2Mn^s$. Since $n = O(N^{1/2})$, the error of the algorithm cannot be less than $O(1/N^{1/2})$. Therefore, we have

THEOREM 2. *The algorithm* (41) *is optimal with respect to the order of the number of operations among all algorithms using $O(n^2)$ data from $I_n(B, b)$.*

One may assume that any algorithm using any N data from $I_n(B, b)$ will yield an error of the order $O(N^{-s/2})$ for the considered class of problems.

In the general case considered here the usual algorithms require an order n^3 of operations, while in the case of equations of convolution type (when the matrix corresponding to the system (22) will be of Toeplitz type), an order n^2 of operations [8].

By repeating the above computation in the case of an equation of convolution type ($A \equiv A_2$), we get, instead of (48),

$$N \asymp n \ln n. \tag{54}$$

This means that the following assertion is true.

THEOREM 3. *In the case of an equation of convolution type and the condition* $M(x,x) \geqslant (A_2^* A_2 x, x) \geqslant m(x,x)$, $m > a$, *the algorithm* (41) *will have a number of operations* N *of the order* $n \ln n$, *which is logarithmically equivalent to the least number of operations possible.*

§4. The case of a singular integral equation

Consider the singular integral equation

$$K\varphi \equiv a(t)\varphi(t) + b(t)S\varphi + \lambda \int_\gamma k(t,\tau)\varphi(\tau)\,d\tau = f(t) \tag{55}$$

under the assumptions that a, b and k (in both variables) have sth derivatives, bounded in absolute value, γ is a circle of unit radius with center at the origin, $a^2 - b^2 \neq 0$, $t \in \gamma$, $|f^{(s)}(t)|^2$ is summable on γ,

$$S\varphi = \frac{1}{\pi i}\int_\gamma \frac{\varphi(\tau)\,d\tau}{\tau - t},$$

the index

$$\kappa = \frac{1}{2\pi}\left[\arg\frac{a-b}{a+b}\right]_\gamma = 0,$$

and that λ is at a distance at least $\delta > 0$ from the nearest eigenvalue of the operator K. It is known [9], [10] that under these assumptions equation (55) has a unique, Hölder continuous solution.

We will consider (55) as a linear equation in the Hilbert space $L_2(\gamma)$ with inner product

$$(f_1, f_2)_{L_2(\gamma)} = \frac{1}{2\pi i}\int_\gamma f_1(t)\cdot\overline{f_2(t)}\,\frac{dt}{t} = \frac{1}{2\pi}\int_0^{2\pi} f_1(t)\overline{f_2(t)}\,dt, \quad t = e^{i\theta}, \quad \left(\|f\| = \sqrt{(f,f)}\right), \tag{56}$$

and with the domain $D(K) \equiv L_2(\gamma)$.

In such a case,

$$K^*\Psi \equiv \bar{a}\psi(t) + \frac{1}{\pi i}\int_\gamma \frac{\bar{b}(\tau)\psi(\tau)}{\tau - t}\,d\tau - \bar{\lambda}\int_\gamma \overline{k(\tau,t)}\,\frac{\psi(t)}{t\tau}\,d\tau. \tag{57}$$

As a system of coordinates $\{e_k\}$ we choose t^k, $k = 0, \pm 1, \pm 2, \ldots$. For this system

$$y_k = Kt^k = \begin{cases} (a+b)t^k + \lambda\int_\gamma k(\tau,t)\tau^k d\tau, & k \geqslant 0, \\ (a-b)t^k + \lambda\int_\gamma k(\tau,t)\tau^k d\tau, & k < 0. \end{cases} \tag{58}$$

It is not hard to check that in this case all the conditions required in the theorems of §3 are fulfilled, provided the input functions a, b, k and f are sufficiently smooth.

It was assumed above that the algorithm uses data from $I_n(B, b)$, i.e. elements of the systems (4) and (20). However, in practice the input often has the form of functions given by tables. If such table-given functions are the kernels of the operator K^*K (or KK^*) at nodes partitioning γ into $2n+1$ equal parts, then in order to obtain the elements (K^*Kt^k, t^j), k, $j = -n, \ldots, n$, or (KK^*t^k, t^j) by the fast Fourier transform (FFT), the number of operations required is of the order $n^2 \ln n$. Specifically, using the FFT we cannot obtain a number of operations of the same order as the number of given data [12], but the logarithmic equivalence will still be observed [13]. Because of the asymptotic properties of the orthonormal systems $\{e_k\}$ [14], the same situation will occur also in the general case for the determination of elements (A^*Ae_k, e_j) or $(AA^*_{e_k,e_j})$.

If the table-given functions are the input elements for equation (55), in order to compute the matrix (K^*Kt^k, t^j) a number of the order of n^3 operations may be required. In order to avoid this unpleasant fact, one can solve (55), e.g. by using the method of moments, which yields the system

$$\sum_{k=-n}^{n} \alpha_k(Kt^k, t^j) = (f, t^j), \qquad j = -n, \ldots, n, \tag{59}$$

that does not require for the determination of its elements a table of values of the iterated kernel.

An approximate determination of the coefficients of the algebraic systems (from table-given functions) produces an additional error of the order $(\ln n)/n^s$ for each such coefficient, which results in an error for the approximate solution of an order at least $(\ln n)/n^{s-2}$. In order to avoid this, one may proceed by means of a piecewise-polynomial approximation of the coefficients (which gets rid of $\ln n$), and by a more refined approximation of the matrix B in the process of proving Theorem 1. Namely, all elements of B should be represented as sums of two terms, $l_{ij} = c_{ij} + d_{ij}$, the first of which corresponds to the compact part of the operator A^*A (or AA^*), and the second to its singular part. In substituting B_k for

B one must then set

$$\begin{cases} b_{ij}^{(k)} = b_{ij}, & i, j = 1, \dots, k, \\ b_{ij}^{(k)} = d_{ij}, & |j - i| \leqslant k, \quad i, j = k + 1, \dots, n, \\ b_{ij}^{(k)} = 0, & |i - j| > k. \end{cases} \tag{60}$$

§5. Some other cases

Applied to equation (55), Fourier transformations reduce equations of convolution type, and in particular Wiener-Hopf equations of the second and (at the expense of a regularizing parameter $\alpha > 0$) first kind [6], [15]. Since the inner products of the pre-images are equal to the inner products of the corresponding transforms [16], the systems of algebraic equations for equations of convolution type and for singular equations are the same. Consequently, the results will also be the same, under assumptions concerning the pre-images guaranteeing the required conditions for singular integral equations (§§3 and 4). The required condition on pre-images is their sufficiently fast decrease at infinity [16].

Quite widespread in potential theory are integral equations with a logarithmic singularity in the kernel [16], [18], [19]. By differentiation these equations can also be reduced to singular integral equations of the form (55).

The main difficulty in approximately solving nonlinear equations consists in finding the initial approximation to the solution of the problem of minimizing the norm of the error (which now is no longer a quadratic function of the unknown parameters), starting with which one can use known iteration methods (the method of simple iteration, Newton's method, and others). This initial approximation can be found by using the results in [20]. By means of these results one can optimally estimate the number of operations in terms of the number of operations required to calculate the residual norm.

If the initial approximation is in the domain of quadratic rate of convergence say, of, Newton's method and obtaining at each iteration the necessary linear approximations of the operator requires an order $n^2 \ln^r n$ of applications, each linear equation being solved as shown in §3, then the final results by the number of operations will still be logarithmically equivalent to the number of input data used.

References

1. K. V. Emel'yanov and A. M. Il'in, *On the number of arithmetical operations required for the approximate solution of a Fredholm integral equation of the second kind*, Zh. Vychisl. Mat. Mat. Fiz. **7** (1967), 905–910; English transl. in USSR Comput. Math. and Math. Phys. **7** (1967).

2. V. V. Ivanov, *Optimal algorithms for the numerical solution of singular integral equations*, Continuum Mechanics and Related Problems of Analysis (N. I. Muskhelishvili Eightieth Birthday Volume), "Nauka", Moscow, 1972, pp. 209–219. (Russian)

3. ______, *On the optimization of methods of numerical solution of singular integral equations*, Materials First All-Union School Theory and Numer. Methods in the Calculation of Shells and Plates, Izdat. Tbiliss. Univ., Tbilisi, 1975, pp. 251–267. (Russian)

4. ______, *Algorithms optimal with respect to accuracy for the approximate solution of operator equations of the first kind*, Zh. Vychisl. Mat. i Mat. Fiz. **15** (1975), 3–11; English transl. in USSR Comput. Math. and Math. Phys. **15** (1975).

5. A. N. Tikhonov and V. Ya. Arsenin, *Methods of solving ill-posed problems*, "Nauka", Moscow, 1974; English transl., Wiley, 1977.

6. V. V. Ivanov, *The theory of approximate methods and its application to the numerical solution of singular integral equations*, "Naukova Dumka", Kiev, 1968; English transl., Noordhoff, 1976.

7. N. M. Korobov, *Number-theoretic methods in approximate analysis*, Fizmatgiz, Moscow, 1963. (Russian)

8. V. V. Voevodin, *Numerical foundations of linear algebra*, "Nauka", Moscow, 1977. (Russian)

9. I. Vecoua [I. N. Vekua], *Sur les équations intégrales linéaires singulières contenant des intégrales au sens de la valeur principale de Cauchy*, C. R. (Dokl.) Acad. Sci. USSR **26** (1940), 327–330.

10. N. I. Muskhelishvili, *Singular integral equations*, 3rd ed., "Nauka", Moscow, 1968; English transl. of 1st ed., Noordhoff, 1953.

11. V. K. Zadiraka, *The fast Fourier transform and some of its applications*, "Znanie", Kiev, 1972. (Ukrainian)

12. Jacques Morgenstern, *Note on a lower bound of the linear complexity of the fast Fourier transform*, J. Assoc. Comput. Mach. **20** (1973), 305–306.

13. N. S. Bakhvalov, *On the properties of optimal methods of solving problems in mathematical physics*, Zh. Vychisl. Mat. i Mat. Fiz. **10** (1970), 555–568; English transl. in USSR Comput. Math. and Math. Phys. **10** (1970).

14. Gabor Szegö, *Orthogonal polynomials*, rev. ed., Colloq. Publ., vol. 23, Amer. Math. Soc., Providence, R. I., 1959.

15. V. V. Ivanov and G. N. Kotenkova, *On an economical method for solving linear operator equations*, Optimization of Computations (Solution of Equations, Numerical Integration, Statistical Data Processing), Preprint No. 40, Inst. Kibernet. Akad. Nauk Ukrain. SSR, Kiev, 1977, pp. 15–27. (Russian)

16. E. C. Titchmarsh, *Introduction to the theory of Fourier integrals*, Clarendon Press, Oxford, 1937.

17. V. Ya. Ivanov and V. P. Il′in, *On the numerical solutions of integral equations in potential theory for model problems*, Preprint No. 33, Seminar on Methods of Numer. and Appl. Math., Computing Center, Siberian Branch Acad. Sci. USSR, Novosibirsk, 1976. (Russian) R. Zh. Mat. **1977** #951078.

18. V. I. Dmitriev and E. V. Zakharov, *The numerical solution of certain Fredholm integral equations of the first kind*, Vychisl. Metody i Programmirovanie Vyp. 10 (1968), 49–54. (Russian)

19. A. N. Tikhonov and V. I. Dmitriev, *A method for calculating the current distribution in a system of linear oscillators and the directivity diagram of this system*, Vychisl. Metody i Programmirovanie Vyp. 10 (1968), 3–8. (Russian)

20. V. V. Ivanov, *Optimal algorithms for the minimization of functions of certain classes*, Kibernetika (Kiev) **1972**, no. 4, 81–94; English transl. in Cybernetics **8** (1972).

Translated by C. EFTIMIU

Amer. Math. Soc. Transl.
(2) Vol. **127**, 1986

Investigation of Properties of Difference Schemes of Through Computation of Second Order of Approximation*

M. YA. IVANOV, V. V. KORETSKIĬ AND N. YA. KUROCHKINA

§1. Introduction

In this article, which is a continuation of a study initiated in [1], we have carried out a qualitative study of properties of difference schemes using techniques of continuous analysis (through computation) for approximations of second order; this study is based on numerical solution of a series of model problems and the analysis of linear and nonlinear equations associated with a differential approximation of the difference schemes under consideration. We study mainly the properties of three widely used difference methods: those of Lax and Wendroff [2] and MacCormack [3], and the "leapfrog" scheme [4]. The relative simplicity of the model problems under consideration allows one to show graphically many characteristic features of the numerical solutions—in particular the oscillating character of the behavior of the computed functions near discontinuities, the rate and extent of "smearing" of discontinuities on a difference mesh, the greater efficiency of the MacCormack scheme compared to that of Lax and Wendroff, the noninvariance of the difference solutions with respect to a Galilean transformation, as well as a number of other properties of second-order schemes.

Following [1], we begin the study with illustrations of several numerical solutions of one-dimensional time-dependent transport equations for $u(x, t)$, a linear equation

$$\frac{\partial u}{\partial t} + a\frac{\partial u}{\partial x} = 0, \qquad a = \text{const} > 0, \tag{1.1}$$

1980 *Mathematics Subject Classification*. Primary 65M99; Secondary 35L65, 35L67, 65D30.
*Translation of Chisl. Metody Mekh. Sploshnoĭ Sredy **11** (1980), no. 2, 41–63; MR **83b**:65091.

and a nonlinear one

$$\frac{\partial u}{\partial t} + \frac{\partial u^2/2}{\partial x} = 0. \tag{1.2}$$

We shall give initial data at $t = 0$ in the form of a "forward" step

$$u = \begin{cases} 1 & \text{for } x < 0, \\ 0.5 & \text{for } x = 0, \\ 0 & \text{for } x > 0, \end{cases} \tag{1.3}$$

a "backward" step

$$u = \begin{cases} 0 & \text{for } x < 0, \\ 0.5 & \text{for } x = 0, \\ 1 & \text{for } x > 0, \end{cases} \tag{1.4}$$

and an isolated wave

$$u(x,0) = ch^{-2}x. \tag{1.5}$$

Along with the calculations of the moving saltus for the nonlinear equation (1.2) with initial data (1.3) we will also present the results of calculating a stationary saltus when the equation

$$\frac{\partial u}{\partial t} + \frac{\partial (u-1)u/2}{\partial X} = 0, \qquad X = x - 0.5t, \tag{1.6}$$

is integrated numerically under initial conditions (1.3).

The accuracy of the numerical results is monitored by comparison with known analytical solutions of the problems and by checking several integral conservation laws. For the solution of the linear equation (1.1) with initial data (1.5) the results satisfy infinitely many conservation laws, which we write in the form

$$I_1 = \int_{-\infty}^{+\infty} u\,dx = \text{const}, \tag{1.7}$$

$$I_2 = \int_{-\infty}^{+\infty} \frac{u^2}{2}\,dx = \text{const}, \tag{1.8}$$

$$I_3 = \int_{-\infty}^{+\infty} \frac{u^3}{3}\,dx = \text{const} \tag{1.9}$$

etc.

For the solution of the nonlinear equation (1.2) with initial data (1.5) only a single law (1.7) is valid. In this case we also have to satisfy the relation

$$\frac{d}{dt}\int_{-\infty}^{+\infty} \frac{u^2}{2}\,dx = \frac{1}{12}(u_+ - u_-)^3, \tag{1.10}$$

where u_- and u_+ are the respective values of u to the left and right of the discontinuity that emerges.

In addition to the direct numerical experiment we will carry out a qualitative analysis of the properties of the through-computation difference schemes on the basis of their linear and nonlinear differential approximations.

§2. The Lax-Wendroff scheme

In this article we examine three widely used explicit difference schemes with second-order accuracy on smooth solutions. The first of these will be the two-step variant of the Lax-Wendroff scheme [2] proposed by Richtmyer.

We shall introduce in xt-space a computational mesh with steps h in the x-coordinate and τ in the t-coordinate indexed as follows: $x = x_j = jh$, $t = t_n = n\tau$, $u_j^n = u(x_j, t_n)$ and so on. As examples here and in what follows we will write out the difference relations only for the nonlinear equation (1.2). The Lax-Wendroff approximation formulas have the form

$$\begin{gathered} \frac{u_{j+1/2}^{n+1/2} - (u_{j+1}^n + u_j^n)/2}{\tau/2} + \frac{(u_{j+1}^n)^2 - (u_j^n)^2}{2h} = 0, \\ \frac{u_{j-1/2}^{n+1/2} - (u_j^n + u_{j-i}^a)/2}{\tau/2} + \frac{(u_j^n)^2 - (u_{j-1}^n)^2}{2h} = 0, \\ \frac{u_j^{n+1} - u_j^n}{\tau} + \frac{(u_{j+1/2}^{n+1/2})^2 - (u_{j-1/2}^{n+1/2})^2}{2h} = 0. \end{gathered} \tag{2.1}$$

The first-step formulas in this scheme coincide with those of Lax [5]. For the stability of computations the condition $|c| \leqslant 1$ must be satisfied, where $c = \tau u/h$ is the Courant number.

Results of solving the problems formulated in the Introduction by the scheme under consideration are shown in Figure 1, in which the mode of presentation is the same as in [1]. Thus the column of graphs on the left corresponds to the linear equation (1.1) when $a = 1$, and the one on the right to the nonlinear equation (1.2). The first, second and third rows depict the results of solving the problems with initial data (1.3), (1.4) and (1.5) for the three moments $t = 1, 2$ and 4, respectively. In the last row are shown the integral characteristics for the solution of the problem of movement of an isolated wave and the results of calculating the stationary saltus by the nonlinear equation (1.6) with initial data (1.3). The calculations are carried out for $h = 0.1$ with two values of the Courant number: the largest possible ($c = 1$) when $\tau = 0.1$, and $c = 0.5$ when $\tau = 0.05$ (the solid and the dashed lines, respectively). In the latter case there is a double "stability provision".

Let us go into more detail about the peculiarities of the numerical solutions. Calculation of the linear equation under discontinuous initial data yields, for practical purposes, an exact solution when $c = 1$ (Figures 1a and 1c). When $c = 0.5$ the numerical solution on the left of the saltus front exhibits the clearly expressed nonmonotonic character and the zone of "smearing" the saltus increases with time. In the nonlinear case the solution is not monotonic near the discontinuities for any admissible value of the Courant number (Figures 1b, 1f,

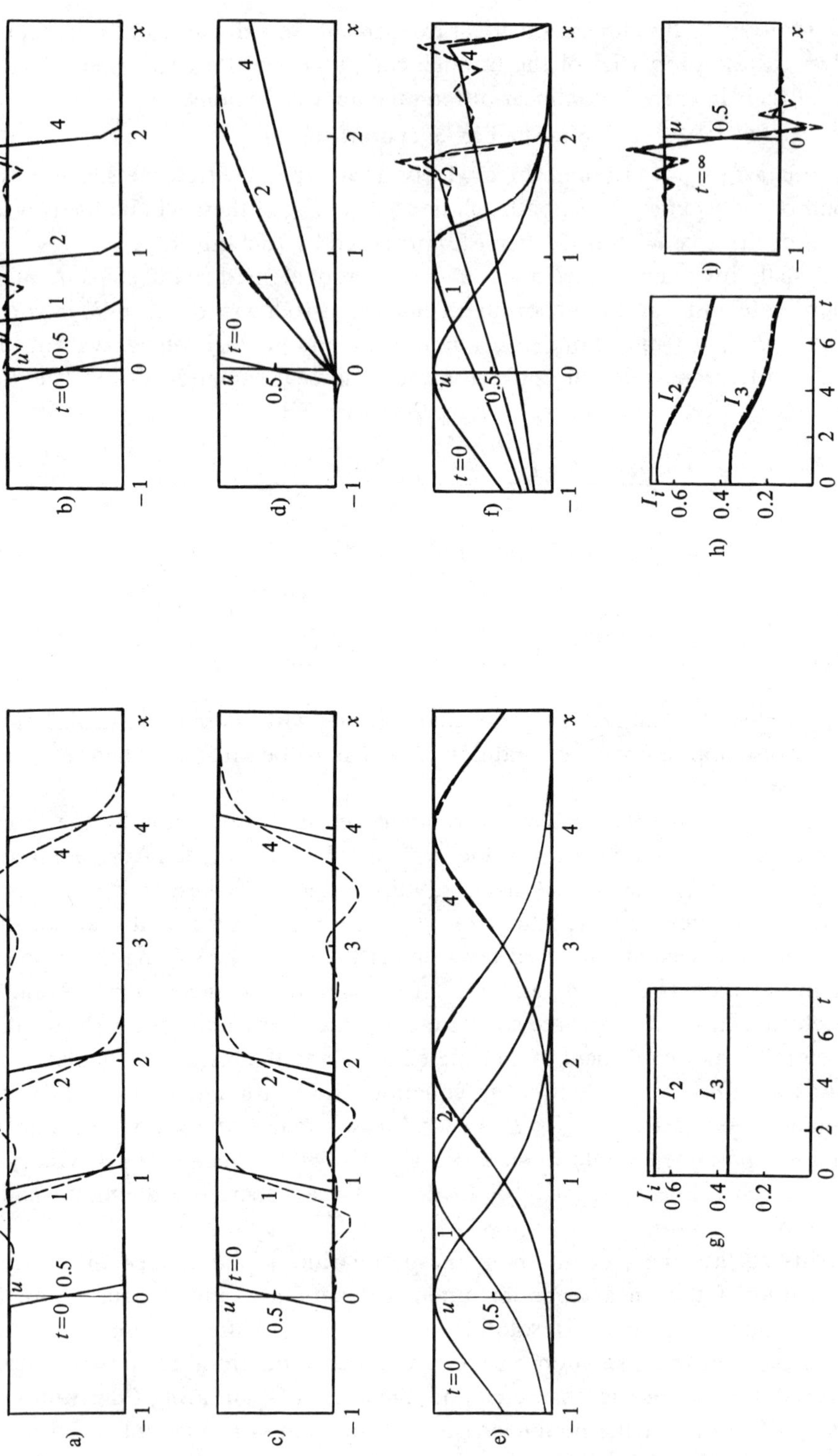

FIGURE 1

1i). For $\tau = 0.05$ the oscillatory nature of the solution in the vicinity of the discontinuities shows up more clearly than for $\tau = 0.1$. The smooth solutions are computed with high accuracy for both $\tau = 0.1$ and $\tau = 0.05$; in the process the integrals (1.7)–(1.9) are preserved to within a few tenths of 1%.

The specifics of the numerical solutions indicated below will be analyzed starting with the linear and nonlinear differential approximations [1]. We shall write down appropriate expressions for the Lax-Wendroff scheme, retaining in the differential approximations as many terms as the purposes of the ensuing presentation require:

$$\frac{\partial u}{\partial t} + a\frac{\partial u}{\partial x} = -\frac{ah^2}{6}(1-c^2)\frac{\partial^3 u}{\partial x^3} - \frac{a^2ch^3}{8}(1-c^2)\frac{\partial^4 u}{\partial x^4}, \tag{2.2}$$

$$\frac{\partial u}{\partial t} + u\frac{\partial u}{\partial x} = -\frac{uh^2}{6}(1-c^2)\frac{\partial^3 u}{\partial x^3} - \frac{h^2}{8}(1-5c^2)\frac{\partial}{\partial x}\left(\frac{\partial u}{\partial x}\right)^2 + \frac{3}{4}\tau u\left(\frac{\partial u}{\partial x}\right)^3 \tag{2.3}$$

§3. The MacCormack scheme

One of the most widely used methods of numerical solution of problems in gas dynamics is that of MacCormack [3]. We shall consider the variant of this method that is used most often. The approximation, as in the Lax-Wendroff scheme, consists of two stages: in the first the intermediate values of the unknown functions are determined, and in the second the final values at the new level being computed. For the nonlinear equation (1.2) the difference formulas have the form

$$\begin{aligned}
&\frac{\bar{u}_j^{n+1} - j_j^n}{\tau} + \frac{\left(u_{j+1}^n\right)^2 - \left(u_j^n\right)^2}{2h} = 0,\\
&\frac{\bar{u}_{j-1}^{n+1} - u_{j-1}^n}{\tau} + \frac{\left(u_j^n\right)^2 - \left(u_{j-1}^n\right)^2}{2h} = 0,\\
&\frac{u_j^{n+1} + \left(\bar{u}_j^{n+1} + u_j^n\right)/2}{\tau/2} + \frac{\left(\bar{u}_j^{n+1}\right)^2 - \left(\bar{u}_{j-1}^{n+1}\right)^2}{2h} = 0.
\end{aligned} \tag{3.1}$$

For the linear equation (1.1) when the coefficient a is constant, the difference relations of the MacCormack scheme on the whole lead to formulas identical with those of Lax and Wendroff. As a consequence, the results for the case of a linear equation computed by the MacCormack scheme coincide with those obtained by the Lax-Wendroff scheme, which we have depicted in Figure 1.

The results of numerically solving the problems with initial data (1.3)–(1.5) for the nonlinear equations (1.2) and (1.6) are shown in Figure 2. The form in which they are represented is the same as that employed in the preceding case for the nonlinear equation (the column of graphs on the right in Figure 1). The main very important distinction, which shows that the MacCormack scheme is more efficient in comparison with the Lax-Wendroff scheme is the more accurate computation in a neighborhood of the discontinuities (smaller oscillations of the parameters). We shall return to these properties when we analyze the appropriate nonlinear differential approximations.

We write down the respective linear and nonlinear differential approximations for the MacCormack scheme:

$$\frac{\partial u}{\partial t} + a\frac{\partial u}{\partial x} = \frac{ah^2}{6}(1-c^2)\frac{\partial^3 u}{\partial x^3} - \frac{a^2ch^3}{8}(1-c^2)\frac{\partial^4 u}{\partial x^4}, \tag{3.2}$$

$$\frac{\partial u}{\partial t} + u\frac{\partial u}{\partial x} = -\frac{uh^2}{6}(1-c^2)\frac{\partial^3 u}{\partial x^3} - \frac{h^2}{4}(1+c-2c^2)\frac{\partial}{\partial x}\left(\frac{\partial u}{\partial x}\right)^2 - \frac{\tau h}{4}(1-2c)\left(\frac{\partial u}{\partial x}\right)^3. \tag{3.3}$$

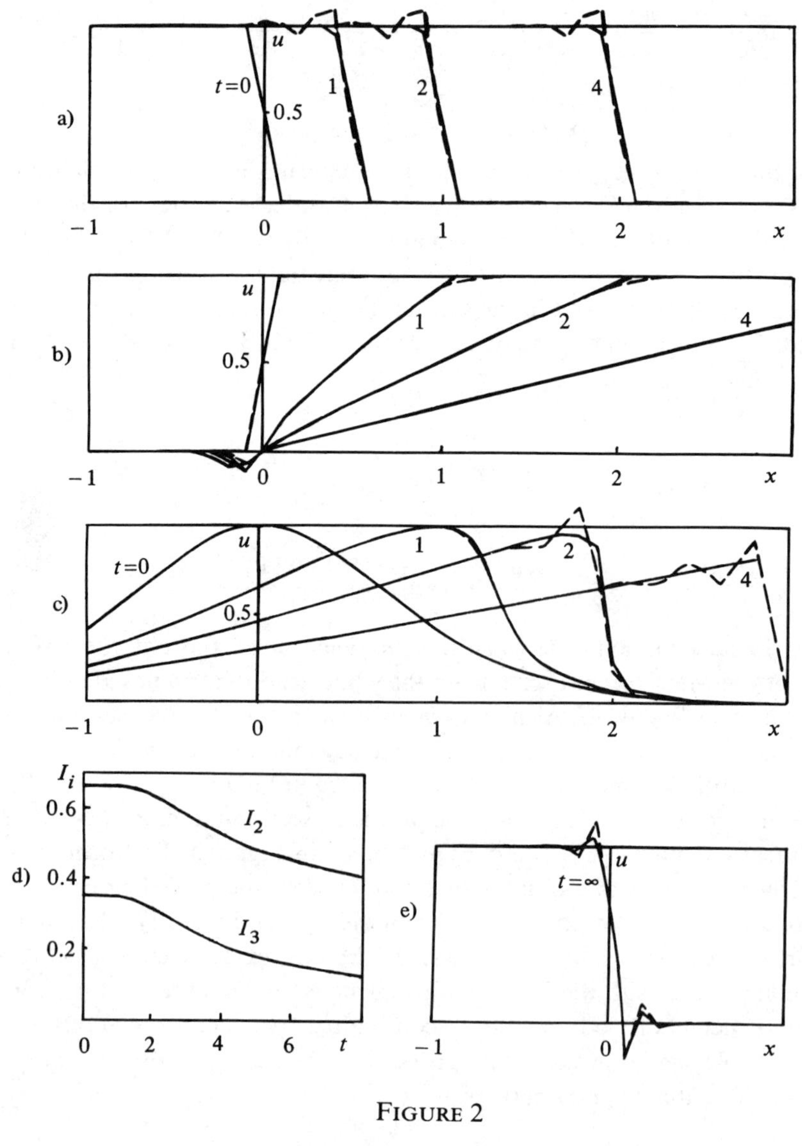

FIGURE 2

It needs to be stressed that the linear differential approximations in the Lax-Wendroff and MacCormack schemes coincide (formulas (2.2) and (3.2)), and that in the nonlinear differential approximations only the coefficients of the two last terms in the first parts of (2.3) and (3.3) are different.

§4. The "leapfrog" scheme

We next consider a three-level divergence scheme of second-order accuracy, which is an analogue of the "cross" scheme presented in a paper of von Neumann and Richtmyer [6] as well as of the "leapfrog" scheme [4]. We write the following difference relation for the nonlinear equation (1.2):

$$\frac{u_j^{n+1} + u_j^{n-1}}{2\tau} = \frac{\left(u_{j+1}^n\right)^2 - \left(u_{j-1}^n\right)^2}{4h} = 0. \tag{4.1}$$

In using this relation it is necessary to assign parameters on the first level in addition to the initial level, and the accuracy of this assignment cannot be less than the accuracy of the "leapfrog" scheme (4.1) itself. In the present article the parameters at the first time level are determined by the Lax-Wendroff formulas. This process is conditionally stable for $|c| \leqslant 1$.

The results of the respective computations are shown in Figure 3. For the linear equation with discontinuous initial data when $c = 1$ the numerical solution has very high accuracy (Figures 3a and 3c). When $c = 0.5$ the character of the solution behind the front of the smeared saltus is oscillational (Figures 3a and 3c) and such that the spread and amplitude of the oscillations exceed the corresponding values in computations by the Lax-Wendroff scheme. A smooth solution of the form (1.5) is also computed with high accuracy in the linear case. Here the errors in determining the integrals I_1, I_2 and I_3 do not exceed 0.04%.

For the nonlinear equation behind the discontinuity front we obtain a strongly oscillating solution (Figures 3b and 3f), which requires either the use of supplementary procedures for smoothing the computed solution or the introduction into the scheme of an artificial viscosity. We observe also that the values of the integrals I_2 and I_3 remain near their initial values and are not diminished when the discontinuity appears; this is inconsistent with the corresponding exact solutions of equation (1.2) with initial data (1.5).

We present the linear and nonlinear differential approximations for the scheme we are analyzing:

$$\frac{\partial u}{\partial t} + a\frac{\partial u}{\partial x} = -\frac{ah^2}{6}(1 - c^2)\frac{\partial^3 u}{\partial x^3} - \frac{ah^4}{120}(1 - c^2)(1 - 9c^2)\frac{\partial^5 u}{\partial x^5}, \tag{4.2}$$

$$\frac{\partial u}{\partial t} + u\frac{\partial u}{\partial x} = -\frac{uh^2}{6}(1 - c^2)\frac{\partial^3 u}{\partial x^3} - \frac{h^2}{4}(1 - 3c^2)\frac{\partial}{\partial x}\left(\frac{\partial u}{\partial x}\right)^2 + \tau^2 u\left(\frac{\partial u}{\partial x}\right)^3. \tag{4.3}$$

We emphasize that the terms containing u_{xxx} in (4.2) and (4.3) are the same as similar terms in the differential approximations of the Lax-Wendroff and MacCormack schemes.

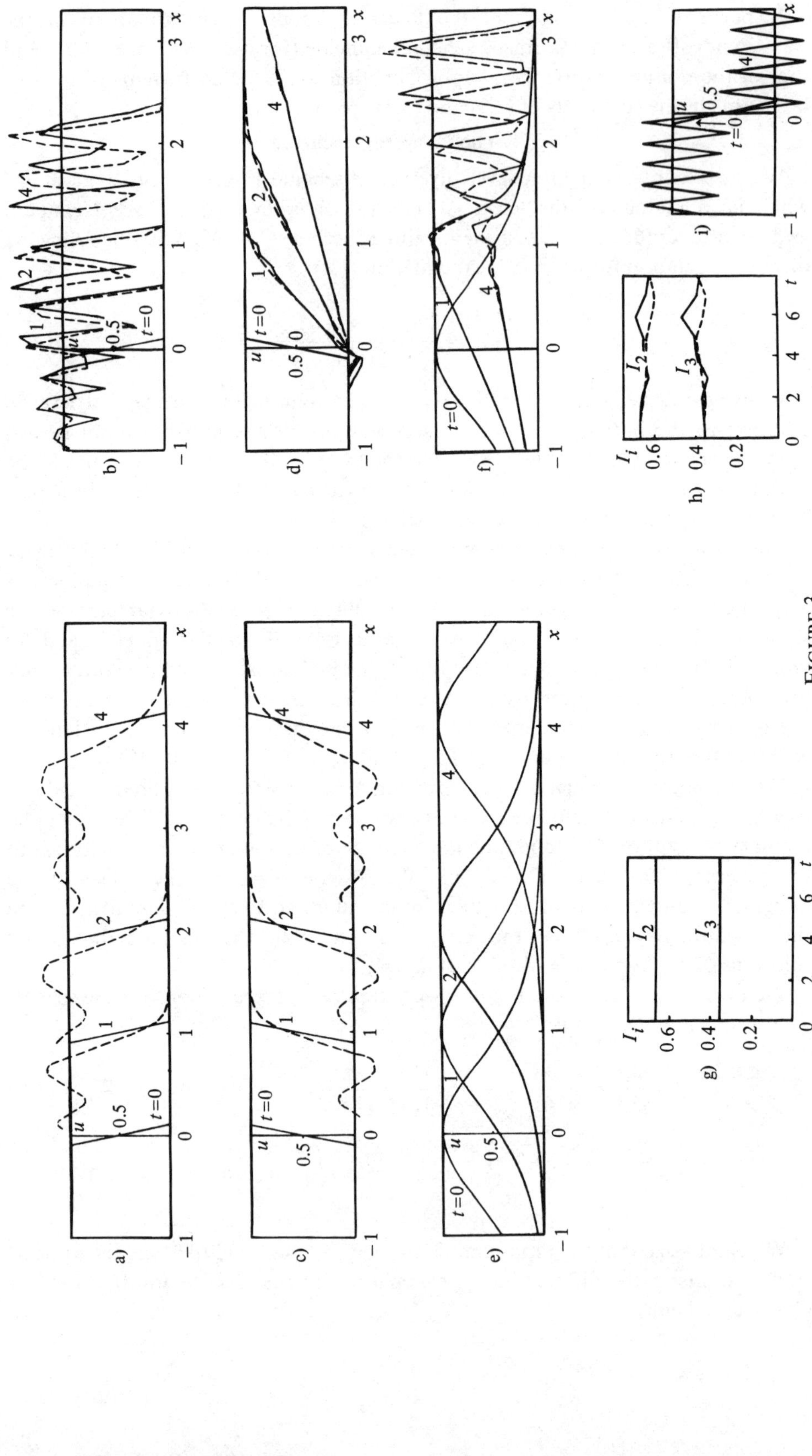

FIGURE 3

With this we conclude our presentation of numerical results on the solution of the simplest model problems by through computation schemes of second-order approximation. Now we shall move on to a qualitative analysis of the main properties of the numerical solutions we have obtained.

§5. Properties of the linear differential approximations

The linear differential approximations of second-order schemes approximating the linear transport equation (1.1) can be written in the following form:

$$\frac{\partial u}{\partial t} + a\frac{\partial u}{\partial x} = \sum_{p=2}^{\infty} \mu_p \frac{\partial^{p+1} u}{\partial x^{p+1}}. \tag{5.1}$$

Having considered the dispersion relation for (5.1), we can carry out the analysis of the dispersive and dissipative properties of the difference schemes. A similar analysis has been made in [7] and [8]. Here we need only mention that in second-order schemes the first dominating term in the approximation error (the first term in the sum on the right side of (5.1)) is the one containing the odd derivative $\partial^3 u/\partial x^3$. By virtue of this, dispersive effects predominate in schemes of second (and any even) order. We shall demonstrate the implications of this for the case of solutions obtained with the aid of through computation schemes of second order for linear and nonlinear transport equations subject to discontinuous initial data. The analysis carried out below supplements the results of the numerical studies set forth in the preceding three sections.

Solution of the linear transport equation with discontinuous initial data can serve as the simplest model of a contact discontinuity. We shall study the properties of the numerical modeling of such a solution by a second-order through computation, starting with the first differential approximations of these schemes, which we write in the form

$$\frac{\partial u}{\partial t} + a\frac{\partial u}{\partial x} = \beta\frac{\partial^3 u}{\partial x^3}. \tag{5.2}$$

This is the linearized Korteweg-de Vries equation. We shall rewrite it in the system of coordinates $X = x - at$:

$$\frac{\partial u}{\partial t} = \beta\frac{\partial^3 u}{\partial X^3} \tag{5.3}$$

and consider the problem with initial data at $t = 0$ in the form of a step

$$u = \begin{cases} 1 & \text{for } X < 0, \\ 0 & \text{for } X > 0. \end{cases} \tag{5.4}$$

Certain conclusions follow at once from the form of equation (5.3) and the initial data (5.4). Thus, as we pass to the dependent variable $y = u - 0.5$, (5.3) keeps its previous form

$$\frac{\partial y}{\partial t} = \beta\frac{\partial^3 u}{\partial X^3}, \tag{5.5}$$

but in place of initial conditions (5.4) we have

$$y = \begin{cases} 0.5 & \text{for } X < 0, \\ -0.5 & \text{for } X > 0. \end{cases} \tag{5.6}$$

By solving (5.5) with initial data in the form of the "forward" step (5.6) we deduce its solution for the "backward" step

$$y = \begin{cases} -0.5 & \text{for } X < 0, \\ 0.5 & \text{for } X > 0, \end{cases} \tag{5.7}$$

since the form of (5.5) does not change when we replace y by $-y$.

This same property is possessed by numerical solutions of the linear transport equation with discontinuous initial data, as shown by the results presented in Figures 1a, 1c, 3a and 3c. Furthermore, the solution of (5.5) with initial data (5.6) for negative values of β can be obtained from the solutions of the same equation with initial data (5.7) for the same (in modulus) positive β by simply replacing X by $-X$.

We turn now to the analysis of the rate of "smearing" of discontinuities of solutions of the linear transport equation as obtained by second-order through-computation difference schemes. The solutions of (5.3) belong to the family of self-similar solutions

$$u = f(z), \qquad z = X(3\beta t)^{-1/3}. \tag{5.8}$$

After substituting (5.8) into (5.3), we obtain the ordinary differential equation

$$f''' + zf' = 0.$$

Hence the region of transition for the solution of (5.2) with discontinuous initial data is close to the line $X = x - at = 0$ and has width proportional to $(\beta t)^{1/3}$. What has just been said explains the "accumulative" mechanism of the smearing of initial discontinuities for the linear equation. Similar properties of numerical solutions are observed also in the case of through computation of contact or tangential discontinuities in problems of gas dynamics.

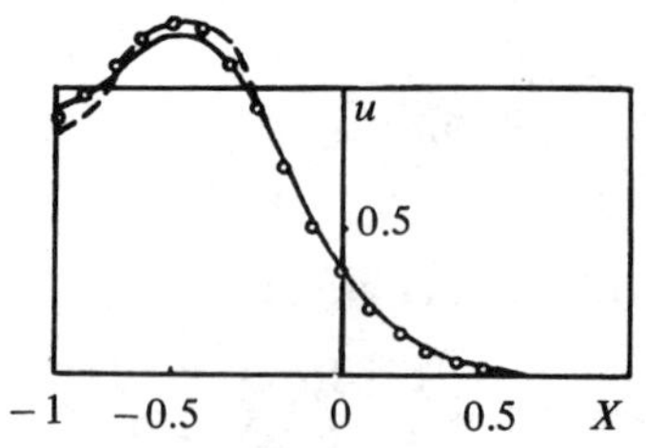

FIGURE 4

Solutions of the linearized Korteweg-de Vries equation are expressed in terms of the Airy integrals—including also those having initial data in the form of a step (see, for example, [9], §13.6). We recall here that when $\beta > 0$ the solution ahead of the principal front of the evolving step is oscillatory, while behind the front it is montonic. When $\beta < 0$ the behavior of the solution near the front is reversed. In Figure 4 the little circles show the solution of (5.3) with initial data (5.4) as a function of the coordinate X at the moment $t = 4$. The magnitude of the coefficient β was chosen equal to the corresponding value from the linear differential approximations of the second-order schemes considered above: $\beta = -ah^2(1 - c^2)/6 = -0.00125$, $h = 0.1$, $\tau = 0.05$, $c = 0.5$. The solid line in Figure 4 is the Lax-Wendroff numerical solution of the linear transport equation (shown earlier in Figure 1a), and the broken line is the numerical solution obtained by the "leapfrog" scheme (Figure 3a) for the same instant $t = 4$. The satisfactory coincidence of the results near the front indicates that the properties of numerical solutions in the regions of large gradients of the parameters are determined mainly by the term βu_{xxx} in the first differential approximation of the second-order schemes. We observe also the difference in the results when we move to the left of the main front of the smeared saltus. There the influence of the next terms in the differential approximations of the schemes becomes noticeable. For the differential approximation of the Lax-Wendroff scheme (2.2) the term in question is $\mu_3 \partial^4 u/\partial x^4$, in which for $c = 0.5$ the value of μ_3 is negative, which causes the presence of a certain mechanism of attenuation ("dissipation") of the oscillations in the solution. For the differential approximation of the "leapfrog" scheme (4.2) the second term in the approximation error, containing the odd derivative $\mu_4 \partial^5 u/\partial x^5$ (for $c = 0.5$ the value of μ_4 is positive), "amplifies" the effects of "dispersion" [7], [8].

§6. Analysis of nonlinear differential approximations

In this section certain properties of second-order difference schemes are demonstrated; they follow from the analysis of nonlinear model equations analogous to the differential approximations (2.3), (3.3) and (4.3) of the difference schemes under consideration when the nonlinear transport equation written in divergence form is solved with their aid. We shall mainly rely on the simplified nonlinear third-order equation

$$\frac{\partial u}{\partial t} + u\frac{\partial u}{\partial x} = \beta\frac{\partial^3 u}{\partial x^2} - \varphi\frac{\partial}{\partial x}\left(\frac{\partial u}{\partial x}\right)^2 + \chi\left(\frac{\partial u}{\partial x}\right)^3, \tag{6.1}$$

assuming that at this stage the coefficients β, φ and χ are constant. In spite of this assumption, the analysis carried out below qualitatively explains several peculiarities of the numerical results, especially the oscillatory behavior of the numerical solutions near discontinuities and the noninvariance of the difference schemes under a Galilean transformation.

We begin with the simplest case, in which only the defining term with third derivative is retained in the right member of (6.1); that is, we consider the widely used Korteweg-de Vries equation

$$\frac{\partial u}{\partial t} + u\frac{\partial u}{\partial x} = \beta\frac{\partial^3 u}{\partial x^3} \tag{6.2}$$

under discontinuous initial data

$$u = \begin{cases} 1 & \text{for } x < 0, \\ 0 & \text{for } x > 0. \end{cases} \tag{6.3}$$

An asymptotic solution of this problem was obtained in [10], and a numerical solution in [11]–[13]. As a result of the decay of the discontinuity an oscillating wave packet is formed, moving ahead of the principal transitional region when $\beta > 0$ and lagging behind it when $\beta < 0$. Moroever, in the course of time more and more new oscillations are formed. In the latter case ($\beta < 0$) the amplitude of the leading oscillation nearest the front attains in finite time the value two, and its velocity becomes equal to a third of the amplitude, i.e., $2/3$. Gradually the amplitudes of the trailing oscillations also take on the value two. When $\beta > 0$ the initial front achieves a velocity equal to $1/3$ in finite time, the amplitude of the oscillation nearest the front becomes equal to -1, and the number of oscillations moving ahead of the front increases with time as well. As the value of the coefficient β grows in magnitude, the distance between two adjacent oscillations lengthens, and the wave packet increases its spread; when β decreases the wave packet shrinks, but the amplitudes of the oscillations retain their previous values. This decay of the initial discontinuity, depending on the presence of dispersive effects, has an entirely different character from that of the nonlinear hyperbolic equation $u_t + uu_x = 0$ and the Burgers equation $u_t + uu_x = \alpha u_{xx}$.

We shall obtain a few quantitative relations for the solution of equation (6.2) with initial data (6.3). Assuming the existence of a solution of stationary form in the coordinate system $X = x - Dt$ and performing the integration with use of the boundary conditions $u'' = 0$ and $u = 1$ as $X \to -\infty$, we arrive at the equation

$$2\beta u'' = (u - D)^2 - (1 - D)^2. \tag{6.4}$$

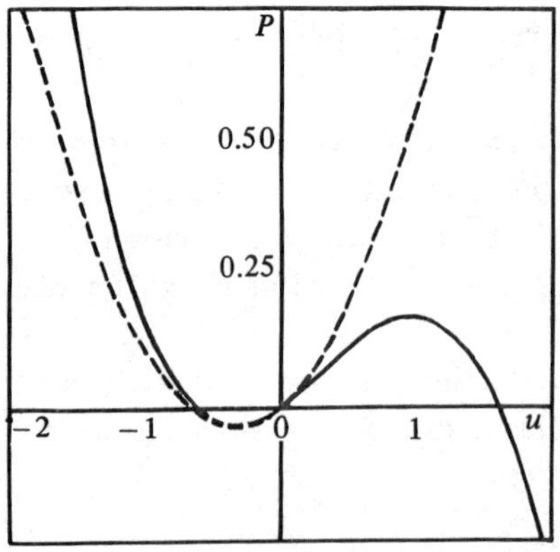

FIGURE 5

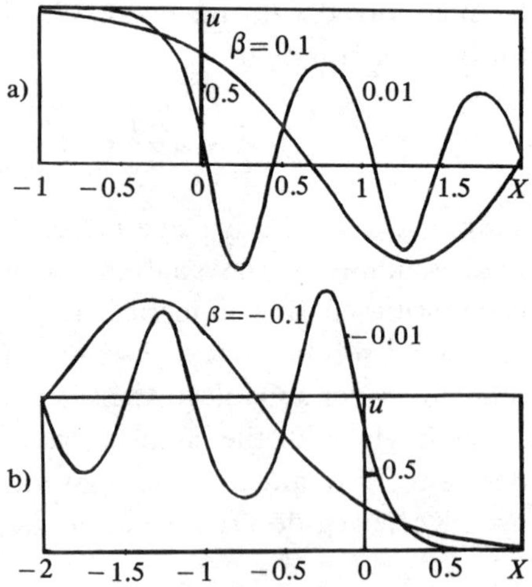

FIGURE 6

This equation is like that of undamped anharmonic oscillations of a single mass point with mass β in a force field determined by the effective potential

$$P(u) = (1/2 - D)u + Du^2/2 - u^3/6. \tag{6.5}$$

The graph of the function $P(u)$ when $D = 1/3$ is shown by the solid line in Figure 5. This function has a local maximum at $u = 1$, at which $P(1) = (2 - 3D)/6$, and a local minimum at $u = 2D - 1$.

Supposing that the differentiation in (6.4) is carried out with respect to dummy time and identifying the function u with the coordinate of the mass point, we arrive at the following mechanical analogue for (6.4). Up to a particular moment the point mass is in unstable equilibrium when $u = 1$, and then it begins to execute undamped anharmonic oscillations in a neighborhood of the stable equilibrium position defined by $u = 2D - 1$. The evolute of the trajectory of the point with respect to coordinate X illustrates the asymptotic behavior of the solution of the problem of decay of an initial discontinuity for the Korteweg-de Vries equation.

From the energy relation for (6.4) we obtain the maximal gradient of the function u. Multiplying (6.4) by u' and integrating, we arrive at the law of conservation of "energy" in the form $\beta(u')^2/2 + P = \text{const}$. On the left is the sum of the kinetic and potential energy of the point mass, which is equal to the greatest value of the potential energy, attained when $u = 1$, i.e., $P(1) = (2 - 3D)/6$. Taking into account that this energy is converted entirely into kinetic when the point passes the equilibrium position, we obtain $u' = \sqrt{(2 - 3D)/3\beta}$. When $D = 1/3$ we have $u' = 1\sqrt{3\beta}$.

Solutions of (6.4) can be expressed in terms of elliptic functions. Near the minimum of the potential energy, $P(u)$ can be approximated by the parabola $P(u) = (1 - D)(u/2 - 2D + 1)u$, shown in Figure 5 by the broken line, and the solution of the linearized equation (6.4) can be obtained in the form

$$u = 2D - 1 + \sqrt{(2 - 3D)/3(1 - D)}\cos\left(\sqrt{(1 - D)/\beta}\,X\right).$$

We now present some illustrations of solutions of the problem we are concerned with. In Figure 6a we have shown solutions of the Korteweg-de Vries equation (6.2) with initial data (6.3) at the instant $t = 1$ for the two positive values $\beta = 10^{-1}$ and 10^{-2}, and in Figure 6b the same for the negative values $\beta = -10^{-1}$ and -10^{-2}.

The properties of the solutions of (6.2) mentioned above are observed in the numerical solutions of nonlinear problems (nonlinear transport equation, equations of nonviscous gas dynamics) by through computation difference schemes of second order of approximation. For example, when the nonlinear transport equation, given in divergence form $u_t + (u^2)_x/2 = 0$, is integrated by the Lax-Wendroff, MacCormack and "leapfrog" difference schemes, the coefficient β in the nonlinear differential approximations of these schemes is equal to $-uh^2(1 - c^2)/6$. If we consider the problem with initial data (6.3), then when the stability condition $|c| \leqslant 1$ is satisfied in computations of a moving saltus for the nonlinear transport equation, we have $u \geqslant 0$. Consequently $\beta \leqslant 0$, leading to generation in the numerical solution of oscillations to the left of the front of the moving saltus (Figures 1b, 2a and 3b). In the computation for the same problem in the coordinate system $X = x - 0.5t$, the corresponding value of β in the three difference schemes will be less than zero when $u > 0.5$ and greater than zero when $u < 0.5$. Therefore oscillations in the numerical solutions will be observed in both the left and right of the saltus front (Figures 1i, 2e and 3i).

The dependence of the sign and magnitude of the coefficient β on the value of u entails the noninvariance of the Lax-Wendroff, MacCormack and "leapfrog" schemes with respect to the Galilean transformation. In the coordinate system $X = x - Dt$, when the nonlinear transport equation with discontinuous initial data (6.3) is solved numerically for $D \geqslant 1$, the oscillations will only be observed ahead of the principal saltus front, and its velocity may be appreciably smaller (when D is significantly greater than one) than the exact solution, which is equal to 0.5. When $D \leqslant 0$ the oscillations will be behind the front of the computed saltus, and the velocity may be substantially larger than 0.5. This property of solutions of the equations of one-dimensional gas dynamics by the Lax-Wendroff scheme (noninvariance under a Galilean transformation) was observed in direct numerical experiments conducted by Yu. I. Shokin (private communication).

In the second differential approximations of the Lax-Wendroff and MacCormack schemes the term with fourth derivative $\partial^4 u/\partial x^4$ is contained, the coefficient having a negative sign. As in the linear case, this guarantees the presence of a certain dissipation mechanism resulting in the smoothing out of the oscillations near the discontinuities. In the differential approximations of the "leapfrog" scheme "dissipative" terms of this kind are missing, implying that the significant oscillations are formed near the discontinuities (Figures 3b, 3f and 3i).

§7. Quadratic scheme viscosity

We now consider the effect on the character of the solution of the additional terms appearing in the nonlinear differential approximations (6.1), beginning with the term

$$\varphi \frac{\partial}{\partial x}\left(\frac{\partial u}{\partial x}\right)^2.$$

We first recall how the discontinuous solution of the nonlinear transport equation $u_t + uu_x = 0$ changes whenever we introduce that term (with a minus sign) on the right-hand side in place of zero. When φ is positive it acts as an artificial quadratic viscosity [14], due to which one obtains, instead of a discontinuity, a transitional region with width of the order of $\sqrt{2\varphi}$. For negative values of φ we cannot construct a continuous smooth solution approximating the discontinuity of the function u. Here a solution of the equation

$$u_t + uu_x + \varphi \frac{\partial}{\partial x}\left(\frac{\partial u}{\partial x}\right)^2 = 0$$

for discontinuous initial data, (6.3) for example, is expressed in terms of the hyperbolic cosine, and the region of its values lies outside the strip $0 < u < 1$.

If we assume the existence in the coordinate system $X = x - Dt$ of a stationary solution describing the variation in the function u from 1 to 0 for the model equation

$$\frac{\partial u}{\partial t} + \frac{1}{2}\frac{\partial u^2}{\partial x} = \beta\frac{\partial^3 u}{\partial x^3} - \varphi\frac{\partial}{\partial x}\left(\frac{\partial u}{\partial x}\right)^2, \tag{7.1}$$

we arrive at an ordinary differential equation; by integrating it once with the conditions $u = u' = u'' = 0$ taken into account as $X \to +\infty$, we get

$$\beta u'' + \varphi(u')^2 + (2D - u)u/2 = 0. \tag{7.2}$$

This equation is analogous to that of anharmonic motion of a point having mass β in a medium with hydraulic resistance when the restoring force depends quadratically on the coordinate u. Linearization of the last term in (7.2) when $u \sim 0$ yields the equation of damped oscillations

$$\beta u'' + \varphi(u')^2 + Du = 0, \tag{7.3}$$

whose properties have been investigated in detail (see, for example, [15]). In particular, when $\beta D > 0$ all solutions of (7.3) are oscillatory; when $\beta D \leqslant 0$ any solution $u(X)$ of (7.3), beginning with some value of the argument, becomes a monotone function having a monotone derivative, and also $u(X) \to 0$ and $u'(X) \to 0$ as $X \to \infty$. A similar linearization for (7.2) can be carried out in the neighborhood of $u \sim 1$, and we can show that the solutions are monotone as $X \to -\infty$ when $\beta(1 - D) \geqslant 0$ and oscillating when $\beta(1 - D) < 0$. It should be emphasized that monotone or oscillating behavior of solutions of (7.3) is determined only by the sign of the coefficient β (when $D > 0$) and is in accord with the behavior of the corresponding solutions of the Korteweg-de Vries equation. The presence of a term with coefficient φ does not change the character of the solution's behavior.

From (7.3) and the condition $u = u' = 0$ as $X \to \infty$ we can obtain the relation

$$(u')^2 = \frac{D\beta}{2\varphi^2}\left(1 + 2\frac{\varphi}{\beta}u - \exp\left(2\frac{\varphi}{\beta}u\right)\right)$$

and the approximate estimate for the solution in a neighborhood of zero: $u \sim \exp(\sqrt{-D/8}\,X)$; from this the same conclusions about the monotonic or oscillating behavior of the solution follow. However, this last estimate does not account for the effects of attenuation in the amplitude of the oscillations when $D/\beta > 0$.

Keeping in mind the material that we have discussed, let us return to the analysis of differential approximations for second-order schemes. The chief parameter determining the qualitative behavior of the solution, as we should have expected, is the coefficient β of the highest derivative u_{xxx} in the first differential approximations of these schemes. When the expressions for this coefficient coincide, as occurs in the Lax-Wendroff, MacCormack and "leapfrog" schemes with $\beta = -uh^2(1 - c^2)/6$, the differences in numerical results depend on the coefficients of the remaining terms in the differential approximations. Thus, in particular, the coefficient φ in the nonlinear differential approximation is $\varphi = h^2(1 - 5c^2)/8$ for the Lax-Wendroff and $\varphi = h^2(1 + c - 2c^2)/4$ for the MacCormack scheme. As the local Courant number $c = u\tau/h$ changes from 0 to 1, the value of φ for the MacCormack scheme varies from $h^2/4$ to 0 and is nonnegative everywhere. This indicates that the term $\varphi(\partial/\partial x)(\partial u/\partial x)^2$, intrinsic to the MacCormack scheme in its differential approximation, acts like the artificial quadratic viscosity added to certain schemes to make possible through computation of discontinuities. At the same time, in the Lax-Wendroff scheme, the function $\varphi(u)$ is of alternating sign and is negative when the Courant numbers c are greater than $\sqrt{1/5}$. Hence, when $\sqrt{1/5} \leqslant c \leqslant 1$ the term $\varphi(\partial/\partial x)(\partial u/\partial x)^2$ does not act like an artificial quadratic viscosity, and this leads to poorer quality of numerical solutions near discontinuities obtained by the Lax-Wendroff scheme as compared with that of MacCormack.

Another very important property of the MacCormack scheme, which we need to keep in mind in through computation of discontinuities, is the fact that in order to ensure that the sign of φ is positive in the whole range $0 \leqslant c \leqslant 1$ when $u < 0$ it is necessary in the first stage of the computation to make use of approximation of the derivative with respect to the spatial coordinate utilizing the left difference—and in the second stage utilizing the right difference—and to change the difference formulae accordingly. This guarantees good efficiency of the MacCormack scheme for any values of u near the saltuses.

An estimation of the effect of the term $\chi(\partial u/\partial x)^3$ in the nonlinear differential approximation of second-order schemes was made by integrating (6.1) according to an implicit symmetric difference scheme of high order of approximation (third with respect to time and fourth with respect to the space coordinate). That estimate showed that for sufficiently small values of χ characteristic for second-order schemes the indicated term does not alter the qualitative behavior of the solution.

In conclusion we underscore the following. In computations of the flow of an ideal gas based on the Euler equations when second-order difference schemes are used, the flow in regions with significant gradients of parameters, and also in zones of separation sometimes observed, is essentially influenced by the quadratic scheme viscosity and effects of "dispersion" connected with the corresponding errors in the approximation of difference schemes. These effects are not only quantitatively but also qualitatively different from the dissipative effects governing the flow of a real gas in zones of separation.

References

1. M. Ya. Ivanov, V. V. Koretskiĭ and N. Ya. Kurochkina, *Investigation of difference schemes of through computation of second order of approximation*, Chisl. Metody Mekh. Sploshnoĭ Sredy **11** (1980), no. 1, 85–114. (Russian)

2. Peter Lax and Burton Wendroff, *Systems of conservation laws*, Comm. Pure Appl. Math. **13** (1960), 217–237.

3. R. W. MacCormack, *The effect of viscosity on hypervelocity impact cratering*, AIAA Paper No. 69-354, 1969.

4. K. Mikoyada, *Introduction to numerical weather forecasting (finite difference methods)*, Numerical Methods of Solving Problems of the Dynamics of the Atmosphere and the Ocean, Gidrometeoizdat, Leningrad, 1968, pp. 7–120. (Russian)

5. Peter D. Lax, *Weak solutions of nonlinear hyperbolic equations and their numerical computation*, Comm. Pure Appl. Math. **7** (1954), 159–163.

6. J. von Neumann and R. D. Richtmyer, *A method for the numerical calculation of hydrodynamic shocks*, J. Appl. Phys. **21** (1950), 232–237.

7. R. F. Warming and B. J. Hyett, *The modified equation approach to the stability and accuracy analysis of finite-difference methods*, J. Computational Phys. **14** (1974), 159–179.

8. Yu. I. Shokin, *On the analysis of the dissipation and dispersion of finite difference schemes*, Chisl. Metody Mekh. Sploshnoĭ Sredy **7** (1976), vyp. 7, 131–141. (Russian)

9. G. B. Whitham, *Linear and nonlinear waves*, Wiley, 1974.

10. A. V. Gurevich and L. P. Pitaevskiĭ, *Decay of initial discontinuity in the Korteweg-de Vries equation*, Pis′ma v Zh. Èksper. Teoret. Fiz. **17** (1973), 268–271; English transl. in JETP Letters **17** (1973).

11. A. C. Vliegenthart, *On finite-difference methods for the Korteweg-de Vries equation*, J. Engrg. Math. **5** (1971), 137–155.

12. Yu. A. Berezin, *On numerical solutions of the Korteweg-de Vries equation*, Chisl. Metody Mekh. Sploshnoĭ Sredy **4** (1973), no. 2, 20–31. (Russian)

13. ______, *Numerical study of nonlinear waves in a rarefied plasma*, "Nauka", Novosibirsk, 1977. (Russian)

14. N. N. Kalitkin, *Numerical methods*, "Nauka", Moscow, 1978. (Russian)

15. Giovanni Sansone, *Equazioni differenziali nel campo reale*. Vol. II, 2nd ed., Zanichelli, Bologna, 1949.

Translated by R. N. GOSS

ABCDEFGHIJ–AMS/AP–89876